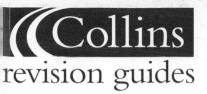
revision guides

TotalRevision

GCSEMaths

Paul Metcalf

Series editor: Jayne de Courcy

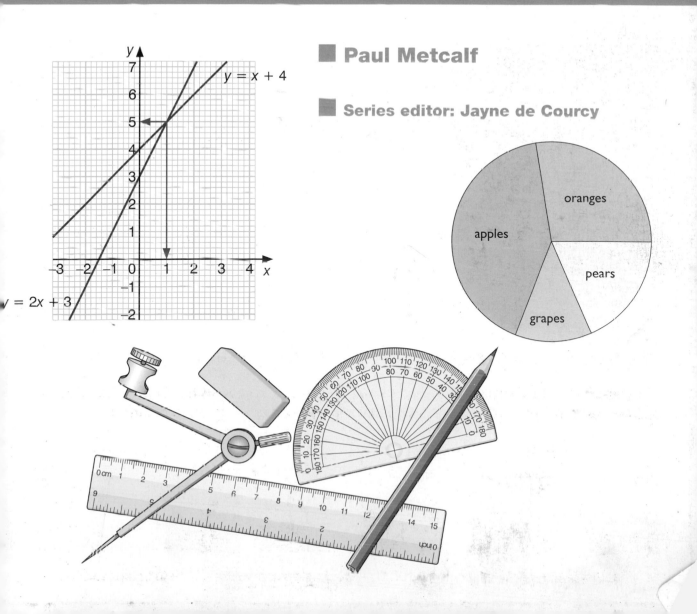

CONTENTS AND REVISION PLANNER

EXAM TIPS

Key comments from Chief Examiners' recent reports
(Comments in blue apply to Higher level only)

Number

Topics tackled well	Topics tackled less well
• odd and even numbers • positive and negative integers • powers of numbers	• the use of fractions • decimals and percentages • decimals and percentages • estimations • recognising prime numbers • reverse percentages • ratio and proportional division • effective use of a calculator • standard form • fractional indices • surds

There was some confusion between "factors" and "multiples". (page 1)

There seemed to be a lot of confusion between 3 significant figures and 3 decimal places. (page 4)

Very few candidates were able to use reverse percentage techniques. (page 20)

A significant number of candidates were let down by poor arithmetic skills, especially the inability to multiply or divide whole numbers accurately or the inability to multiply or divide decimals accurately. (page 24)

A small but significant number of candidates use calculators set on radians or gradients. (page 26)

A significant number of candidates were unable to rationalise the denominator of a fraction with surds included. (page 32)

Although candidates knew how to convert recurring decimals to fractions, they were not able to give any proof or explanation. (page 33)

There was considerable confusion with fractional indices. Numbers with negative indices were often interpreted as negative numbers. (page 34)

Algebra

Topics tackled well	Topics tackled less well
• basic algebra • number patterns • substitution into a formula • expanding brackets	• changing the subject • interpreting graphs • inequalities • simultaneous equations • algebraic fractions • simultaneous equations, one linear, one non-linear • completing the square

There were some very good responses ... but candidates often penalised themselves with incomplete simplifications or incorrect notation in algebra. (page 41)

Centres need to ensure that candidates are aware of the difference between giving a formula and giving an expression. Many candidate omitted the y =...., and could not therefore be given full marks. (pages 41, 47)

The graph plotting was quite accurate although some candidates lost marks. When plotting points or drawing graphs, the usual level of tolerance is ±2 small square or ± 1 mm. (page 51)

Simultaneous equations continued to cause problems for many candidates. Despite the clear instruction not to use trial and improvement, which is penalised, a large number of candidates insisted on ignoring this. (page 65)

Whilst a large number of candidates were familiar with the elimination method for solving simultaneous equations, they often lost marks through arithmetic errors involving negative signs. (page 66)

Few candidates understood the concept of completing the square and many did not attempt this question. (page 74)

It was disappointing to see many errors in use of the quadratic formula. Common errors included incorrect substitution of negative numbers or dividing only part of the numerator by 2. (page 75)

Shape, space and measures

Topics tackled well	Topics tackled less well
• area and volume • using Pythagoras theorem • vectors	• knowing geometrical terms • constructions • imperial and metric units • transformations • congruency and similarity • volume of a prism • elevation • scale factors for areas and volumes • enlargement with negative scale factors

The terms "obtuse" and "reflex" were not well known. (page 82)

Candidates could identify the pentagon, but it was common to see incorrect answers such as 'quadrilateral', 'hexagon' and 'house shape'. (spage 85)

Candidates should be encouraged to use a sharp pencil for their drawing work. (page 88)

Few candidates new how to construct the perpendicular bisector accurately. (page 90)

A surprising number of candidates did not know that 180° was the three figure bearing for South. (page 92)

Very few candidates could convert between metric and imperial units. (page 94)

Fully correct answers were extremely rare and candidates forgot to include all the details of the rotation, including the angle of rotation or the centre of rotation. (page 100)

The questions stated that the diagrams were "not drawn accurately" to indicate that measuring will lead to an incorrect answer. However, many candidates ignored this and used a protractor, despite the instruction to calculate the value of the angle. (page 104)

Some candidates had no idea which trigonometric ratio to use. (page 112)

Candidates could find the volume of a cuboid but were unable to multiply the three dimensions. (page 117)

Angles in a semi circle were not recognised or incorrectly used. (page 120)

Candidates frequently had problems in rearranging the sine rule correctly. (page 130)

A large number of candidates used the cosine formula incorrectly by carrying out the calculation in the wrong order. (page 130)

Handling data

Many candidates completed the pie chart correctly but a significant number lost accuracy marks by drawing sectors which were outside the 2° tolerance. (page 147)

A number of candidates misread their protractors. (page 147)

Very few candidates seemed to know how to calculate a mean, with many giving the mode or the median. (page 151)

A large number of candidates did not know how to calculate the mid-points of the class intervals, especially where the classes had different widths. (page 152)

Cumulative frequency graphs were not plotted at the upper class boundary. (page 157)

Candidates did not always understand a line of best fit and thought that the line must pass through the origin. (page 160)

There are a significant number of candidates using incorrect notation (e.g. 1 out of 6, 1 in 6 or 1:6) for probability questions. (page 162)

Few candidates were able to provide the correct definitions for a random and a stratified sample. (page 169)

General comments

■ Candidates should be reminded to read through each question carefully, as they frequently lost marks through failing to give their answer in the form requested.

■ Candidates should not round until the end of the question. They should give their answers either as exact answers or correct to three significant figures unless otherwise instructed.

■ The standard of written English was often poor, with the spelling of mathematical words particularly weak, and candidates unable to express themselves well when an explanation was required.

■ There are still too many candidates not showing any working out, which means that the available method marks cannot be earned when the final answer is wrong.

■ It is encouraging to see that candidates are now remembering to give the units of their answers where these are not given.

■ Not all candidates had access to a ruler, protractor or pair of compasses. In fact, it was more common to see any circle drawing question attempted freehand!

ABOUT THIS BOOK

Exams are about much more than just repeating memorised facts. This book has been planned to make your revision as active and effective as possible.

How?

- by breaking down the content into manageable chunks (Revision Sessions)

- by testing your understanding at every step of the way (Check Yourself Questions)

- by providing separate sessions to help you gain the very top grades (Higher Tier sessions)

- by giving as much guidance as possible (**Worked examples and Hint boxes**)

- by giving you invaluable examiner's guidance about exam technique (Exam Practice)

REVISION SESSION

Revision Sessions

- This book covers all the topics for GCSE Maths **Intermediate** and **Higher** tier exams. **All** the revision sessions are relevant for **Higher** tier students. Those sessions that are **not** relevant for **Intermediate** tier students are marked by a yellow strip down the side of each page.

- Each topic is divided into a number of **short revision sessions**. You should be able to read through each of these in about half an hour. That is the maximum amount of time that you should spend on revising without taking a short break.

CHECK YOURSELF QUESTIONS

- At the end of each revision session there are some **Check Yourself Questions**. By trying these questions, you will immediately find out whether you have understood and remembered what you have read in the revision session. **Answers** are at the back of the book, along with **extra hints and guidance**.

- If you manage to answer all the Check Yourself questions for a session correctly, then you can confidently tick off this topic in the box provided in the Contents list. If not, you will need to tick the 'Revise again' box to remind yourself to return to this topic later in your revision programme.

Higher Tier sessions

■ Some topics include material that is relevant only to students revising for the Higher Tier exam. These sessions are indicated by a yellow strip along the outside of the page. They are also marked in a similar way in the Contents list.

Worked Examples and Hint Boxes

Throughout the book there are lots of worked examples. The author has chosen these carefully. They will remind you how to tackle each aspect of Maths. If there is something you haven't understood properly, the worked examples will give you **all the help you need**.

The Hint boxes placed beside many of the worked examples give you **extra guidance and support** with your revision.

Calculator and Non-calculator Methods

Your exam papers will consist of one calculator paper and one non-calculator paper. You must therefore spend time revising and practising both **calculator and non-calculator methods** for answering questions. The ▲ symbol alongside worked examples and questions means that these should be answered by non-calculator methods.

Exam Practice

This unit will give you **invaluable guidance on how to answer exam questions well**.

There are also some **typical exam questions** for you to try to answer. Answers are given at the back of the book so that you can check your own answers against them. There are also examiner's hints highlighting how to achieve **full marks**.

Reading through these will give you a very clear idea of what you need to do in order to score **full marks** when answering questions in your GCSE Maths exam.

Working through this section will give you an excellent grounding in exam technique. If you feel you want further exam practice, look at **Do Brilliantly GCSE Maths**, also published by Collins Educational.

BOOST YOUR GRADE

About your Maths course

All GCSE Mathematics specifications (syllabuses) have to conform to the requirements of the Mathematics National Curriculum. For this reason, the content and assessment of the different GCSE examinations are very similar. There may be small differences in the way that coursework is assessed or in the inclusion of a mental arithmetic or aural test.

Tiers of entry

The Mathematics examination is offered at three tiers. The available grades are as follows:

Tier of entry

A*	A	B	C	Higher
B	C	D	E	Intermediate
D	E	F	G	Foundation

Your teacher will advise you on which tier is best for you.

Assessment

The assessment for all GCSE Mathematics examinations is divided into four assessment objectives. The assessment objectives are weighted as follows:

Using and applying mathematics	20%
Number and algebra	40%
Shape, space and measures	20%
Handling data	20%

Part of the assessment objectives for *Using and applying mathematics* and for *Handling data* are assessed through coursework involving investigational and statistical work.

What this book covers

This book covers all the content of the assessment objectives tested on Intermediate and Higher tier written papers. It is divided into four sections covering:
Number Revision sessions 1–20, pages 1–36
Algebra Revision sessions 1–20, pages 37–81
Shape, space and measures Revision sessions 1–21, pages 82–142
Handling data Revision sessions 1–9, pages 143–

Each section includes work covering the Intermediate tier and the Higher tier of entry. If you are entered for the Intermediate tier of entry, then you do not need to cover the work for the Higher tier which is collected at the end of each section. Each of the revision sessions that relate to work for the Higher tier has HIGHER written below the session numbers, and there is a yellow band down the edge of each page.

Written papers

The GCSE Mathematics specification (syllabus) is assessed on the basis of terminal examination papers taken at the end of the course (linear courses) or modular examination papers taken throughout the course (modular courses). The written papers assess the full range of grades for the tier of entry and are divided into calculator and non-calculator papers. The papers consist of questions of varying lengths and you will be required to answer all the questions on the paper.

For the written papers you will be expected to have the following equipment:

- pens
- sharp pencil, pencil sharpener and eraser
- ruler, protractor and compasses
- a scientific calculator.

Formula sheets

Each of the written papers also includes a formula sheet with a variety of formulae provided. Details of these formulae are given in this book in the relevant chapters.

UNIT 1: NUMBER

The number system

≫ What are integers?

An **integer** is a whole number. It may be positive (⁺1, ⁺2, ⁺3, ...) or negative (⁻1, ⁻2, ⁻3, ...).

≫ What are multiples?

The **multiples** of a number are the products in the multiplication tables.

■ Multiples of 5 are 5, 10, 15, 20, 25, 30, ...

■ Multiples of 6 are 6, 12, 18, 24, 30, 36, ...

The **least common multiple** of two or more numbers is the least (lowest) multiple that is common to all of the given numbers.

Worked example

Find the least common multiple of 3 and 5.

Multiples of 3:	3, 6, 9, 12, 15, 18, 21, 24, 27, 30, ...
Multiples of 5:	5, 10, 15, 20, 25, 30, 35, 40, 45, ...
Common multiples of 3 and 5:	15, 30, 45, 60, ...

The least common multiple of 3 and 5 is 15.

≫ What are factors?

The **factors** (divisors) of a number are numbers that divide *exactly* into that number (i.e. **without a remainder**). The number 1 and the number itself are *always* factors of the given number.

■ Factors of 8 are 1, 2, 4 and 8 because each of these divides exactly into 8.

■ Factors of 13 are 1 and 13 because each of these divides exactly into 13.

A number with exactly two factors is a **prime number**. The numbers 2, 3, 5, 7, 11, 13, 17, ... are all prime numbers.

Remember that the number **1 is not a prime number** because it has only one factor.

The **highest common factor** of two or more numbers is the highest factor which is common to all of the given numbers.

Worked example

Find the highest common factor of 16 and 24.

Factors of 16:	1, 2, 4, 8 and 16
Factors of 24:	1, 2, 3, 4, 6, 8, 12 and 24
Common factors of 16 and 24:	1, 2, 4 and 8

The highest common factor of 16 and 24 is 8.

+ HINTS

▸ Factors of a number come in pairs. For example:
$$24 = 1 \times 24$$
$$= 2 \times 12$$
$$= 3 \times 8$$
$$= 4 \times 6$$
Factors of 24 are:
1, 2, 3, 4, 6, 8, 12, 24.

▸ Square numbers such as 16 always have an odd number of factors as one of the factor pairs is made up of the same factor, multiplied by itself.

≫ What are prime factors?

A **prime factor** is a factor that is also a prime number.

All numbers can be written as a **product of prime factors**.

■ The number 15 can be written as 3×5 where 3 and 5 are prime factors.

■ The number 90 can be written as $2 \times 3 \times 3 \times 5$ where 2, 3 and 5 are prime factors.

You can find the prime factors of a number by rewriting it in **factor pairs**. Carry on doing this until all the factor pairs are prime numbers.

Worked example

Write 90 as a product of its prime factors.

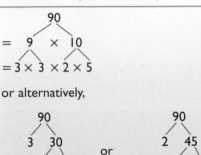

$$90$$
$$= 9 \times 10$$
$$= 3 \times 3 \times 2 \times 5$$

or alternatively,

or

So $90 = 3 \times 3 \times 2 \times 5$ or $90 = 2 \times 3^2 \times 5$

? CHECK YOURSELF QUESTIONS

Q1 From the numbers in the grid on the right write down:
  a) a multiple of 7
  b) a factor of 10
  c) a factor of 51
  d) a square number bigger than 10
  e) a prime number bigger than 16
  f) a prime number that is even
  g) a number that is a multiple of 3 and also a multiple of 7.

1	2	3	4	5
6	7	8	9	10
11	12	13	14	15
16	17	18	19	20
21	22	23	24	25

Q2 Write 264 as a product of its prime factors.

Answers are on page 184.

Directed numbers

What are directed numbers?

A directed number has a + or − sign in front of it.

You will see directed numbers used in temperature scales, where negative numbers show temperatures below freezing.

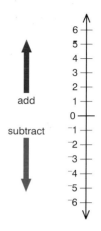

Adding and subtracting directed numbers

To add or subtract directed numbers, find the starting position, then move up or down the number line. On a horizontal number line, move right or left.

Worked example

$2 + 3 = {}^+5$	Start at ${}^+2$ and move up 3 places to ${}^+5$.
$4 - 7 = {}^-3$	Start at ${}^+4$ and move down 7 places to ${}^-3$.
${}^-3 + 6 = {}^+3$	Start at ${}^-3$ and move up 6 places to ${}^+3$.
${}^-1 - 4 = {}^-5$	Start at ${}^-1$ and move down 4 places to ${}^-5$.

Multiplying and dividing directed numbers

To multiply or divide directed numbers, multiply or divide the numbers and then attach the sign according to these rules.

- If the signs are the same, the answer will be positive.
- If the signs are opposite, the answer will be negative.

So:

$$+ \times + = + \qquad - \times - = +$$
$$+ \times - = - \qquad - \times + = -$$

and:

$$+ \div + = + \qquad - \div - = +$$
$$+ \div - = - \qquad - \div + = -$$

+ HINT

▸ When two signs appear together (e.g. $5 - {}^-4$), replace them with one sign, using these rules.

$+ \; +$ gives $+$
$+ \; -$ gives $-$
$- \; +$ gives $-$
$- \; -$ gives $+$
$(-1) + (-2)$ is the same as ${}^-1 - 2$ as $+ \; -$ gives $-$.
$(+2) - ({}^-3)$ is the same as ${}^+2 + 3$ as $- \; -$ gives $+$.

Worked example

${}^+3 \times {}^-4 = {}^-12$	As $+ \times - = -$.
${}^-6 \div {}^+2 = {}^-3$	As $- \div + = -$.
${}^-4 \div {}^-8 = {}^+\frac{1}{2}$	As $- \div - = +$ and $\frac{4}{8} = \frac{1}{2}$.

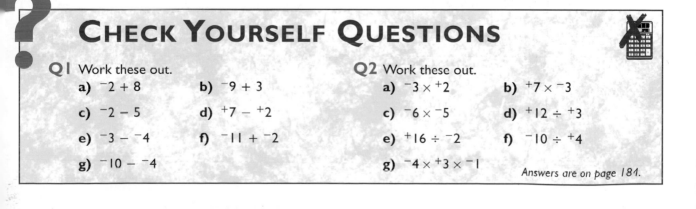

CHECK YOURSELF QUESTIONS

Q1 Work these out.

a) ${}^-2 + 8$ b) ${}^-9 + 3$

c) ${}^-2 - 5$ d) ${}^+7 - {}^+2$

e) ${}^-3 - {}^-4$ f) ${}^-11 + {}^-2$

g) ${}^-10 - {}^-4$

Q2 Work these out.

a) ${}^-3 \times {}^+2$ b) ${}^+7 \times {}^-3$

c) ${}^-6 \times {}^-5$ d) ${}^+12 \div {}^+3$

e) ${}^+16 \div {}^-2$ f) ${}^-10 \div {}^+4$

g) ${}^-4 \times {}^+3 \times {}^-1$

Answers are on page 181.

Rounding – significant figures and decimal places

≫ Why use rounding?

Quite often, an **approximate** answer is acceptable. **Rounding** gives approximate answers. Rounding is very common for numbers in everyday life, for example:

- populations are often expressed to the nearest million

- the number of people attending a pop concert may be expressed to the nearest thousand

- inflation may be expressed to the nearest whole number, or the nearest tenth of a percentage.

≫ Significant figures

Any number can be rounded to a given number of **significant figures** (written s.f.).

Use the following rules.

- Count along the digits to the required number of significant figures.

- Look at the **next digit** (to the right) in the number.
  If its value is less than 5, leave the digit before it as it is.
  If its value is 5 or more, increase the digit before it by 1.

- Replace all the digits to the right, but before the decimal point, by zeros, to keep the number at its correct size.

  Digits to the right after the decimal point can just be left out.

Worked example
Round 7638.462 to the number of significant figures shown.

6 s.f. 7638.462 = 7638.46 (6 s.f.)

5 s.f. 7638.462 = 7638.5 (5 s.f.)

4 s.f. 7638.462 = 7638 (4 s.f.)

3 s.f. 7638.462 = 7640 (3 s.f.)   Fill with 0s to keep the number at its correct size.

2 s.f. 7638.462 = 7600 (2 s.f.)   Fill with 0s to keep the number at its correct size.

1 s.f. 7638.462 = 8000 (1 s.f.)   Fill with 0s to keep the number at its correct size.

≫ Decimal places

Any number can be rounded to a given number of **decimal places** (written d.p.).

Use the following rules.

- Count along the digits to the required number of decimal places.

- Look at the **next digit** (to the right) in the number.
  If its value is less than 5 leave the digit before it as it is.
  If its value is 5 or more, increase the digit before it by 1.

- Leave out all the digits to the right.

➕ *HINT*

▸ Numbers to the left of the decimal point are not affected by this rounding process.

Worked example

Round 6.427 509 3 to the number of decimal places shown.

6 d.p.	6.427 509 3 = 6.427 509 (6 d.p.)
5 d.p.	6.427 509 3 = 6.427 51 (5 d.p.)
4 d.p.	6.427 509 3 = 6.4275 (4 d.p.)
3 d.p.	6.427 509 3 = 6.428 (3 d.p.)
2 d.p.	6.427 509 3 = 6.43 (2 d.p.)
1 d.p.	6.427 509 3 = 6.4 (1 d.p.)

❓ CHECK YOURSELF QUESTIONS

Q1 Write each of the following correct to 3 s.f., 2 s.f. and 1 s.f.
  a) 174.9   **b)** 699.06

Q2 Write 0.8006 correct to 3 d.p., 2 d.p. and 1 d.p.

Answers are on page 184.

Powers, roots and reciprocals

≫ Powers and roots

Multiplying a number by itself one or more times gives a **power** of the first number. The base number is a **root** of the power.

- $5 \times 5 = 5^2 = 25$
  5^2 is 5 raised to the **power** of 2 or 5 **squared**.
  5 is the **square root** of 25.

≫ What are squares and cubes?

A **square number** is formed by multiplying a number by itself.

- The square of 8 is $8 \times 8 = 64$ so 64 is a square number.

A **cube number** is formed when another number is multiplied by itself and then multiplied by itself again.

- The cube of 5 is $5 \times 5 \times 5 = 125$ so 125 is a cube number.

≫ Square roots and cube roots

The **square root** of a number is the number which, when squared, gives the first number.

- The square root of 36 is 6, because $6 \times 6 = 36$.
  The square root of 36 is also $^-6$, because $^-6 \times ^-6 = 36$.

The sign $\sqrt[2]{}$ or $\sqrt{}$ is used to denote the square root so $\sqrt{36} = 6$ or $\sqrt{36} = ^-6$, which you usually write as $\sqrt{36} = \pm 6$.

Remember that **the square root of a number may be positive or negative**.

You can use the $\boxed{\sqrt{}}$ key on your calculator to find the square root of a number, but it will only give the positive square root.

The **cube root** of a number such as 27 is the number which, when cubed, gives the first number (27).

- The cube root of 27 is 3 because $3 \times 3 \times 3 = 27$.

The sign $\sqrt[3]{}$ is used to denote the cube root so $\sqrt[3]{27} = 3$.

You can find a cube root on a calculator if you have a key marked $\boxed{\sqrt[3]{}}$.

≫ Reciprocals

To find the **reciprocal** of a number you divide 1 by that number.

You can find the reciprocal of any non-zero number by converting the number to a fraction and turning the fraction upside-down.

■ The reciprocal of $\frac{2}{3}$ is $\frac{3}{2}$ and the reciprocal of 10 is $\frac{1}{10}$.

With a calculator you can find the reciprocal of a number by using the $\boxed{1/x}$ or $\boxed{x^{1}}$ key.

You may need to use the $\boxed{\text{INV}}$ or $\boxed{\text{2ndF}}$ key with it.

? CHECK YOURSELF QUESTIONS

Q1 Work out $\sqrt{36}$, $\sqrt{10}$.

Q2 Work out $\sqrt[3]{4096}$, $\sqrt[3]{-10}$.

Q3 Work out the reciprocals of $\frac{3}{4}$, 15, $1\frac{1}{5}$.

Answers are on page 185.

Positive, negative and zero indices

≫ Rules for indices

When you multiply a number by itself, use the following shorthand.

- $5 \times 5 = 5^2$ Say, '5 to the power 2 (or 5 squared)'.
- $5 \times 5 \times 5 = 5^3$ Say, '5 to the power 3 (or 5 cubed)'.

In general, you write the shorthand like this:

5^4 ← power
     ← base

and you say it as 'five to the power four'.

The power (or **index**) tells you how many times to multiply the base number.

5^4 tells you to multiply together 4 (the power or index) 'lots' of 5 (the base number).

- $5^4 = 5 \times 5 \times 5 \times 5$

Similarly 4^7 tells you to multiply together 7 'lots' of 4.

- $4^7 = 4 \times 4 \times 4 \times 4 \times 4 \times 4 \times 4$

> **HINT**
>
> ▸ You are multiplying the base number by itself, not by the index.

≫ How to multiply numbers with indices

You can multiply numbers with indices like this.

$3^4 = 3 \times 3 \times 3 \times 3$ $3^5 = 3 \times 3 \times 3 \times 3 \times 3$

So $3^4 \times 3^5 = (3 \times 3 \times 3 \times 3) \times (3 \times 3 \times 3 \times 3 \times 3)$

$= 3 \times 3 \times 3 \times 3 \times 3 \times 3 \times 3 \times 3 \times 3$

$= 3^9$

To multiply two numbers with indices **when their bases are the same** you just add their indices.

- $3^4 \times 3^5 = 3^{4+5} = 3^9$ and $12^4 \times 12^6 = 12^{4+6} = 12^{10}$

> **HINT**
>
> ▸ In general: $a^m \times a^n = a^{m+n}$

≫ How to divide numbers with indices

You can divide numbers with indices like this.

$$6^7 \div 6^4 = \frac{6 \times 6 \times 6 \times \cancel{6} \times \cancel{6} \times \cancel{6} \times \cancel{6}}{\cancel{6} \times \cancel{6} \times \cancel{6} \times \cancel{6}} = 6 \times 6 \times 6 = 6^3$$

To divide two numbers with indices **when their bases are the same** you just subtract their indices.

> **HINT**
>
> ▸ In general: $a^m \div a^n = a^{m-n}$

- $6^7 \div 6^4 = 6^{7-4} = 6^3$ and $15^9 \div 15^3 = 15^{9-3} = 15^6$

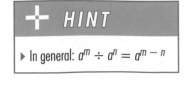

≫ Negative powers

From the above it follows that:

$$7^4 \div 7^6 = 7^{4-6} = 7^{-2} \quad \text{and} \quad 7^4 \div 7^6 = \frac{\cancel{\frac{1}{7}} \times \cancel{\frac{1}{7}} \times \cancel{\frac{1}{7}} \times \cancel{\frac{1}{7}}}{7 \times 7 \times \cancel{\frac{7}{1}} \times \cancel{\frac{7}{1}} \times \cancel{\frac{7}{1}} \times \cancel{\frac{7}{1}}} = \frac{1}{7^2}$$

So $7^{-2} = \frac{1}{7^2}$

≫ Zero powers

Using the same ideas as before:

$$5^4 \div 5^4 = 5^{4-4} = 5^0 \quad \text{and} \quad 5^4 \div 5^4 = \frac{\cancel{\frac{1}{5}} \times \cancel{\frac{1}{5}} \times \cancel{\frac{1}{5}} \times \cancel{\frac{1}{5}}}{\cancel{\frac{5}{1}} \times \cancel{\frac{5}{1}} \times \cancel{\frac{5}{1}} \times \cancel{\frac{5}{1}}} = 1$$

$5^0 = 1$, $6^0 = 1$, $100^0 = 1$, $645. \, 321^0 = 1$ and so on.

? CHECK YOURSELF QUESTIONS

Q1 Find the value of the following.

   a) 9^3 **b)** 4^{-2} **c)** 6^1

Q2 Calculate these, giving your answers in index form where possible.

   a) $3^{11} \times 3^{12}$ **b)** $8^6 \div 8^4$

   c) $13^4 \div 13^4$ **d)** $4^3 \times 5^2$

Answers are on page 185.

Standard form involving positive and negative indices

≫ What is standard form?

Standard form is a short way of writing very large and very small numbers. You must always write standard form numbers in the form:

$A \times 10^n$

where A is a number from 1 to 10 and n is an integer which tells you how many times to multiply by 10 (if n is positive) or divide by 10 (if n is negative).

A may take the value 1, but it must *always* be less than 10.

≫ Very large numbers

Worked example

Write 35 000 in standard form.

Place the decimal point so A is a number from 1 to 10 and find n.

3.5 0 0 0   so $A = 3.5$ and $n = 4$.

$35\,000 = 3.5 \times 10^4$

≫ Very small numbers

Worked example

Write 0.000 000 478 in standard form.

Place the decimal point so A is a number from 1 to 10 and find n.

0 0 0 0 0 0 0 4.78   so $A = 4.78$ and $n = {}^-7$.

$0.000\,000\,478 = 4.78 \times 10^{-7}$

+ **HINT**

▸ In standard form, n is positive for large numbers (e.g. $3.5 \times 10^4 = 35\,000$) and n is negative for small numbers (e.g. $4.78 \times 10^{-7} = 0.000\,000\,478$).

≫ Adding and subtracting numbers in standard form

To add (or subtract) numbers in standard form **when the powers are the same**, proceed as shown in the next example.

Worked example

Work out $(4.18 \times 10^{11}) + (3.22 \times 10^{11})$.

$(4.18 \times 10^{11}) + (3.22 \times 10^{11}) = (4.18 + 3.22) \times 10^{11}$
$= 7.4 \times 10^{11}$

To add (or subtract) numbers in standard form **when the powers are *not* the same** you need to **convert the numbers to ordinary form.**

Worked example

Work out $(8.42 \times 10^6) + (6 \times 10^7)$.

$(8.42 \times 10^6) + (6 \times 10^7) = 8\,420\,000 + 60\,000\,000$	Converting to ordinary form.
$= 68\,420\,000$	
$= 6.842 \times 10^7$	Converting back to standard form.

HINT

▸ Your calculator will deal with numbers in standard form, if you use the EXP or EE key.

≫ Multiplying and dividing numbers in standard form

Use the rules of indices to multiply (or divide) numbers in standard form. (See Revision session 5, *Positive, negative and zero indices*.)

Worked example

Work out $(8.5 \times 10^3) \times (4.2 \times 10^7)$.

$(8.5 \times 10^3) \times (4.2 \times 10^7) = (8.5 \times 4.2) \times (10^3 \times 10^7)$ Collecting powers of 10.

$= 35.7 \times 10^{3+7}$

Using the rules of indices where $a^m \times a^n = a^{m+n}$

$= 35.7 \times 10^{10}$

$= 3.57 \times 10^{11}$

Writing 35.7 as 3.57×10^1 to get the final answer in standard form.

Worked example

Work out $(6.3 \times 10^5) \div (2.1 \times 10^8)$.

$(6.3 \times 10^5) \div (2.1 \times 10^8) = (6.3 \div 2.1) \times (10^5 \div 10^8)$

Collecting powers of 10 together.

$= 3 \times 10^{5-8}$

Using the rules of indices where $a^m \div a^n = a^{m-n}$.

$= 3 \times 10^{-3}$

CHECK YOURSELF QUESTIONS

Q1 a) The distance from the Earth to the Moon is 250 000 miles. Write this number in standard form.

b) The weight of a hydrogen atom is given as 0.000 000 000 000 000 000 001 67 milligrams. Write this number in standard form.

Q2 Work these out without a calculator, then use a calculator to check your answers.

a) $(2.69 \times 10^5) - (1.5 \times 10^5)$
b) $(4.31 \times 10^{-4}) + (3.5 \times 10^{-4})$
c) $(6 \times 10^3) \times (4 \times 10^{-2})$
d) $(3 \times 10^{11}) \div (6 \times 10^4)$

Q3 Light travels at 2.998×10^8 m/s. How far does it travel in one year?

Give your answer in metres, using standard index form.

Answers are on page 186.

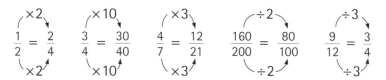

≫ The parts of a fraction

The top part of a fraction is called the **numerator** and the bottom part of a fraction is called the **denominator**.

≫ Equivalent fractions

You can find equivalent fractions by multiplying or dividing the numerator and denominator by the same number.

$$\overset{\times 2}{\frac{1}{2}} = \underset{\times 2}{\frac{2}{4}} \qquad \overset{\times 10}{\frac{3}{4}} = \underset{\times 10}{\frac{30}{40}} \qquad \overset{\times 3}{\frac{4}{7}} = \underset{\times 3}{\frac{12}{21}} \qquad \overset{\div 2}{\frac{160}{200}} = \underset{\div 2}{\frac{80}{100}} \qquad \overset{\div 3}{\frac{9}{12}} = \underset{\div 3}{\frac{3}{4}}$$

≫ Cancelling fractions

You can express a fraction in its **lowest terms** or **simplest form** by making the numerator and the denominator as small as possible. Both numerator and denominator must be integers. The process of reducing fractions to their lowest terms is called **cancelling down** or **simplifying**.

≫ One number as a fraction of another

To find one number as a fraction of another, write the numbers as a fraction, with the first number as the numerator and the second as the denominator.

Worked example

Write 55p as a fraction of 80p.

55p as a fraction of 80p $= \frac{55}{80} = \frac{11}{16}$ so 55p is $\frac{11}{16}$ of 80p.

Worked example

Write 4 mm as a fraction of 8 cm.

You must first ensure that the units are the same.

8 cm = 80 mm

4 mm as a fraction of 80 mm $= \frac{4}{80} = \frac{1}{20}$
So 4 mm is $\frac{1}{20}$ of 8 cm.

≫ Addition and subtraction

To add (or subtract) fractions, make sure they have the same denominator.

Worked example

Add $\frac{2}{7} + \frac{4}{7}$.

$$\frac{2}{7} + \frac{4}{7} = \frac{6}{7}$$

Subtract $\frac{7}{8} - \frac{1}{5}$.

$$\frac{7}{8} - \frac{1}{5} = \frac{35}{40} - \frac{8}{40} = \frac{27}{40}$$

Writing both fractions with a denominator of 40.

> ### ✚ HINT
> ▸ To find the common denominator of two fractions with different denominators you find the least common multiple of the two denominators.

≫ Mixed numbers

A mixed number is made up of a whole number part and a fractional part, such as $1\frac{1}{5}$ or $5\frac{13}{20}$.

Any mixed number can be converted to an **improper fraction** (or **top-heavy fraction**).

- $1\frac{1}{5} = 1 + \frac{1}{5} = \frac{5}{5} + \frac{1}{5} = \frac{6}{5}$ where $1 = \frac{5}{5}$

- $5\frac{13}{20} = 5 + \frac{13}{20} = \frac{100}{20} + \frac{13}{20} = \frac{113}{20}$ where $5 = \frac{100}{20}$

Worked example

Add $1\frac{1}{5} + 5\frac{13}{20}$.

$$1\frac{1}{5} + 5\frac{13}{20} = \frac{6}{5} + \frac{113}{20} = \frac{24}{20} + \frac{113}{20}$$

$$= \frac{137}{20} = 6\frac{17}{20}$$

Converting to improper fractions and writing both fractions with a denominator of 20.

Rewriting as a mixed number.

$$\frac{6}{5} \xrightarrow{\times 4} \frac{24}{20}$$

≫ Multiplication of fractions

To multiply fractions, multiply the numerators and multiply the denominators.

Worked example

Work out $\frac{4}{7} \times \frac{2}{11}$.

$$\frac{4}{7} \times \frac{2}{11} = \frac{4 \times 2}{7 \times 11}$$

$$= \frac{8}{77}$$

When working with mixed numbers you *must* convert to improper fractions first.

Worked example

Work out $1\frac{1}{5} \times 6\frac{2}{3}$.

$$1\frac{1}{5} \times 6\frac{2}{3} = \frac{6}{5} \times \frac{20}{3}$$ Converting to top-heavy fractions.

$$= \frac{6 \times 20}{5 \times 3}$$ Multiplying the numerators and multiplying the denominators.

$$= \frac{120}{15} = 8$$

> ### ✚ HINT
> ▸ Where possible, cancel the fractions.
>
> $$\frac{\cancel{6}^{2} \times \cancel{20}^{4}}{\cancel{5}_{1} \times \cancel{3}_{1}} = \frac{2 \times 4}{1 \times 1} = 8$$

▸ Alternatively, you can cancel the fractions.

$$\frac{2 \times \cancel{100}^{20}}{\cancel{5} \times 1}_{1} = \frac{2 \times 20}{1 \times 1}$$

$$= \frac{40}{1} = 40$$

Worked example

Find $\frac{2}{5}$ of 100.

$\frac{2}{5}$ of $100 = \frac{2}{5} \times \frac{100}{1}$ Writing 100 as a top-heavy fraction.

$\qquad = \frac{\cancel{200}^{40}}{\cancel{5}_{1}}$ Cancelling.

$\qquad = 40$

≫ Division of fractions

To divide one fraction by another, multiply the first fraction by the **reciprocal** of the second fraction.

Worked example

Work out $\frac{3}{7} \div \frac{1}{7}$.

$\frac{3}{7} \div \frac{1}{7} = \frac{3}{\cancel{7}_{1}} \times \frac{\cancel{7}^{1}}{1}$ Multiplying by the reciprocal and cancelling the fractions.

$\qquad = 3$

≫ Changing fractions to decimals

Use division to change a fraction to a decimal.

Worked example

Change $\frac{1}{4}$ to a decimal.

$\qquad \frac{1}{4} = 1 \div 4$

$\qquad \quad = 0.25$

Worked example

Change $\frac{4}{15}$ to a decimal.

$\qquad \frac{4}{15} = 4 \div 15$

$\qquad \quad = 0.266\,666\,6 \ldots$

Note that the decimal in the example above carries on for ever. It is a **recurring decimal**.

- In the decimal $0.2\dot{6}$ the dot over the 6 tells you that the number carries on infinitely.

If a group of numbers repeats infinitely then you can use two dots to show the repeating numbers.

- $0.\dot{3}\dot{5} = 0.353\,535\,35 \ldots$
- $6.4\dot{1}\dot{7} = 6.417\,171\,717 \ldots$
- $3.\dot{2}0\dot{1} = 3.201\,201\,201 \ldots$
- $11.60\dot{2}\,5\dot{3} = 11.602\,532\,532\,53 \ldots$

≫ Changing decimals to fractions

You can change a decimal to a fraction by considering place value as follows.

Worked example

Change 0.459 to a fraction.

$0.459 = 0$ units and 4 tenths and 5 hundredths and 9 thousandths

$$0.459 = 0 + \frac{4}{10} + \frac{5}{100} + \frac{9}{1000}$$

$$= \frac{400}{1000} + \frac{50}{1000} + \frac{9}{1000}$$

Writing the fractions with a common denominator of 1000.

$$= \frac{459}{1000}$$

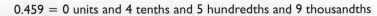

? CHECK YOURSELF QUESTIONS

Q1 Work out $3\frac{1}{4} - 1\frac{1}{5}$.

Q2 Work out $\frac{3}{4} \times \frac{2}{5}$.

Q3 Work out $4\frac{4}{5} \div 1\frac{1}{15}$.

Q4 Change 0.162 to a fraction.

Answers are on page 186.

Percentages

≫ What is a percentage?

A percentage is a number of parts per 100.

- 50% means 50 parts per 100 or $\frac{50}{100} = \frac{1}{2}$ in its lowest terms.

≫ Changing percentages, fractions and decimals

PERCENTAGES TO FRACTIONS

To change a percentage to a fraction, divide by 100.

Worked example

Change **68%** to a fraction.

$$68\% = \frac{68}{100}$$

$$= \frac{17}{25} \quad \text{Cancelling down to the lowest terms.}$$

Worked example

Change $45\frac{1}{2}\%$ to a fraction.

$$45\frac{1}{2}\% = \frac{45\frac{1}{2}}{100}$$

$$= \frac{90}{200} \quad \text{Multiplying top and bottom by 2 to give integers on the top and bottom.}$$

PERCENTAGES TO DECIMALS

To change a percentage to a decimal, divide by 100.

Worked example

Change **68%** to a decimal.

$$68\% = 68 \div 100$$
$$= 0.68$$

Worked example

Change $72\frac{3}{4}\%$ to a decimal.

$$72\frac{3}{4}\% = \frac{72\frac{3}{4}}{100}$$

$$= \frac{291}{400} \quad \text{Multiplying top and bottom by 4 to remove the fraction on the top.}$$

$$= 291 \div 400$$
$$= 0.7275$$

+ **HINT**

▶ As $72\frac{3}{4} = 72.75$, you could just divide 72.75 by 100 to get the same result.

FRACTIONS TO PERCENTAGES

To change a fraction to a percentage, multiply by 100.

Worked example

Change $\frac{1}{4}$ to a percentage.

$$\frac{1}{4} = \frac{1}{4} \times 100\%$$
$$= 25\%$$

To change a decimal to a percentage, multiply by 100.

Worked example

Convert 0.2 to a percentage.

$$0.2 = 0.2 \times 100\%$$
$$= 20\%$$

Worked example

Change 0.005 to a percentage.

$$0.005 = 0.005 \times 100\%$$
$$= 0.5\%$$

≫ Comparing percentages, fractions and decimals

To compare and order percentages, fractions and decimals, change them all into percentages.

Worked example

Place the following in order of size, starting with the smallest.

$65\%, \frac{3}{5}, 0.62, 63.5\%, \frac{3}{4}, 0.7$

65%	65%
$\frac{3}{5} = \frac{3}{5} \times 100\% = 60\%$	60%
$0.62 = 0.62 \times 100\% = 62\%$	62%
63.5%	63.5%
$\frac{3}{4} = \frac{3}{4} \times 100\% = 75\%$	75%
$0.7 = 0.7 \times 100\% = 70\%$	70%

So the order is $\frac{3}{5}, 0.62, 63.5\%, 65\%, 0.7, \frac{3}{4}$ (smallest to highest).

≫ Expressing one number as a percentage of another

To express one number as a percentage of another, write the first number as a fraction of the second and convert the fraction to a percentage.

Worked example

Write 55p as a percentage of 88p.

55p as a fraction of 88p $= \frac{55}{88}$.

$\frac{55}{88} = \frac{55}{88} \times 100\%$ Converting the fraction to a percentage.

$\quad = 62.5\%$

Worked example

Write 2 feet as a percentage of 5 yards.

First you must ensure that the units are the same.

5 yards = 15 feet   As 1 yard = 3 feet

So the problem becomes 'write 2 feet as a percentage of 15 feet'.

2 feet as a fraction of 15 feet $= \frac{2}{15}$.

$\frac{2}{15} = \frac{2}{15} \times 100\%$ Converting the fraction to a percentage.

$\quad = 13.333333...\%$

$\quad = 13.3\%$ (3 s.f.)

≫ Finding a percentage of an amount

To find a percentage of an amount, find 1% of the amount and then multiply to get the required amount.

Worked example

Calculate 50% of £72.

$$1\% \text{ of } £72 = £\frac{72}{100}$$
$$= £0.72$$
$$50\% \text{ of } £72 = 50 \times £0.72$$
$$= £36$$

Worked example

An investment valued at £2000 shows an increase of 6% one year. What is the new value of the investment?

To find 6% of £2000, first find 1%.

$$1\% \text{ of } £2000 = £20$$
$$\text{So } 6\% \text{ of } £2000 = 6 \times £20$$
$$= £120$$

The new value of the investment is £2000 + £120 = £2120

> **HINT**
>
> ▸ 106% is equivalent to a multiplier of 1.06 ($\frac{106}{100}$).

An alternative method uses the fact that after a 6% increase, the new amount will be 100% of the original amount + 6% of the original amount.

100% + 6% = 106% of the original amount

The new value of the investment is 106% of £2000.

$$1\% \text{ of } £2000 = £20$$
$$106\% \text{ of } £2000 = 106 \times £20$$
$$= £2120 \text{ (as before)}$$
$$\text{or } 1.06 \times £2000 = £2120$$

Worked example

A caravan valued at £8000 depreciates by 9% each year. What is the value of the caravan after:

a) one year   **b)** two years?

After a depreciation of 9%, the new amount

$$= 100\% \text{ of the original amount} - 9\% \text{ of the original amount}$$
$$= 91\% \text{ of the original amount}$$

> **HINT**
>
> ▸ 91% is equivalent to a multiplier of 0.91 ($\frac{91}{100}$).

a) After one year, the value of the caravan is 91% of £8000.

$$1\% \text{ of } £8000 = £80$$
$$91\% \text{ of } £8000 = 91 \times £80$$
$$= £7280$$
$$\text{or} \quad 0.91 \times £8000 = £7280$$

b) After the second year, the value of the caravan is 91% of £7280.

$$1\% \text{ of } £7280 = £72.80$$
$$91\% \text{ of } £7280 = 91 \times £72.80$$
$$= £6624.80$$
$$\text{or} \quad 0.91 \times £7280 = £6624.80$$

≫ Percentage change

To work out the percentage change, work out the increase or decrease and add or subtract it from the original amount.

$$\text{Percentage change} = \frac{\text{change}}{\text{original amount}} \times 100\%$$

where the change might be an increase, decrease, profit, loss, error, etc.

Worked example

A company produces 78 000 parts one year and 79 950 the following year. Calculate the percentage increase.

$$\text{Percentage increase} = \frac{\text{increase}}{\text{original amount}} \times 100\%$$

$$\text{Increase} = 79\,950 - 78\,000 = 1950$$

$$\text{Percentage increase} = \frac{1950}{78\,000} \times 100\% = 2.5\%$$

Worked example

A car valued at £3200 is sold for £2950. What is the percentage loss?

$$\text{Percentage loss} = \frac{\text{decrease}}{\text{original amount}} \times 100\%$$

$$\text{Loss} = £3200 - £2950 = £250$$

$$\text{Percentage loss} = \frac{250}{3200} \times 100\% = 7.8125\%$$

$$= 7.81\% \text{ or } 7.8\% \text{ to an appropriate degree of accuracy.}$$

? CHECK YOURSELF QUESTIONS

Q1 The highest percentage of people unemployed in Great Britain was in 1933. Out of 13 million people available for work, 3 million were unemployed.

What percentage were unemployed? Give your answer to the nearest whole number.

Q2 The price of a bathroom suite is advertised as £2800. A discount of 5.5% is agreed for a speedy sale. What is the final cost of the bathroom suite?

Q3 A weighing machine records a weight of 6 kg when the actual weight is 6.005 kg. What is the percentage error on the actual weight?

Answers are on page 187.

Reverse percentages

≫ What is a reverse percentage?

When an amount is increased or decreased by a given percentage, the result is a new amount. Use the technique of reverse percentages to:

- find the original amount after a percentage change
- find the percentage by which the amount was changed.

≫ Using reverse percentages

Worked example

A CD system costs £282 including VAT at $17\frac{1}{2}$%. What is the cost of the CD system without the VAT?

117.5% of the cost of the CD system = £282

£282 represents 117.5% (100% + 17.5%) of the cost of the CD system.

$$1\% \text{ of the cost of the CD system} = £\frac{282}{117.5}$$

$$= £2.40$$

$$100\% \text{ of the cost of the CD system} = 100 \times £2.40$$

$$= £240$$

The CD system costs £240 without the VAT.

Worked example

A washing machine is advertised at £335.75 after a price reduction of 15%. What was the original price of the washing machine?

£335.75 represents 85% of the original price (100% − 15%).

$$\text{So} \quad 85\% \text{ of the original price} = £335.75$$

$$1\% \text{ of the original price} = £\frac{335.75}{85}$$

$$= £3.95$$

$$100\% \text{ of the original price} = 100 \times £3.95$$

$$= £395$$

The original price of the washing machine was £395.

? CHECK YOURSELF QUESTIONS

Q1 The price of a holiday is reduced by 5% to £361. What was the original cost of the holiday?

Q2 A car is sold for £5225 after a depreciation of 45% of the original purchase price. Calculate the original purchase price of the car.

Answers are on page 186.

≫ Why use estimates?

Estimation and approximation are important elements of the non-calculator examination paper. You will be required to give an estimation by rounding numbers to convenient approximations, usually one significant figure.

≫ Rounding

Worked example

Estimate the value of $\frac{40.68 + 61.2}{9.96 \times 5.13}$.

Rounding each of these figures to 1 s.f. gives $\frac{40 + 60}{10 \times 5} = \frac{100}{50} = 2$

Using a calculator, the actual answer is 1.993 940 7... so the answer is a good approximation.

Worked example

Estimate the value of $\frac{\sqrt{98.6}}{2.13^3 + 1.88}$.

Rounding these figures to 1 d.p. gives $\frac{\sqrt{100}}{2^3 + 2} = \frac{10}{8 + 2} = \frac{10}{10} - 1$

Using a calculator, the actual answer is 0.860 195 765 ...

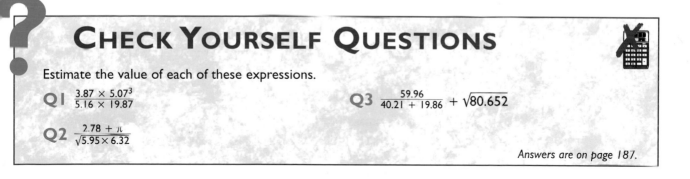

CHECK YOURSELF QUESTIONS

Estimate the value of each of these expressions.

Q1 $\frac{3.87 \times 5.07^3}{5.16 \times 19.87}$

Q2 $\frac{2.78 + \pi}{\sqrt{5.95 \times 6.32}}$

Q3 $\frac{59.96}{40.21 + 19.86} + \sqrt{80.652}$

Answers are on page 187.

Ratios and proportional division

≫ How do you use ratios?

You can use a ratio to compare one quantity to another quantity.
Ratios work in a similar way to fractions.

Worked example

In a box there are 12 lemons and 16 oranges. Write this as a ratio comparing the number of lemons to the number of oranges.

The ratio of the number of lemons to the number of oranges is 12 to 16.

You write this as $12:16$.

The order is important in ratios.

The ratio of the number of oranges to the number of lemons is 16 to 12 or $16:12$.

≫ Equivalent ratios

Equivalent ratios are ratios that are equal to each other.

The following ratios are all equivalent to 2 : 5.

$2:5 = 4:10 = 6:15 = 8:20 = \dots$

Equivalent ratios can be found by multiplying or dividing both sides of the ratio by the same number. You can use this method to obtain the ratio in a form where both sides are integers.

- $1:2 = 2:4$
- $3:7 = 15:35$
- $16:20 = 4:5$
- $0.5:5 = 1:10$
- $\frac{3}{11}:\frac{4}{11} = 3:4$
- $12.5:15 = 5:6$

≫ Cancelling ratios

A ratio can be expressed in its **simplest form** or **lowest terms** by making both sides of the ratio as small as possible. Remember that both sides of the ratio must be integers.

Worked example

Write each of these ratios in its simplest form.

a) $10:15$ b) $121:44$ c) $4:\frac{1}{4}$ d) $2.5:0.5$

a) $10:15 = 2:3$

Dividing both sides by 5.

b) $121:44 = 11:4$

Dividing both sides by 11.

c) $4:\frac{1}{4} = 16:1$

Multiplying both sides by 4 to get integer values.

d) $2.5:0.5 = 5:1$

Multiplying both sides by 2 to get integer values.

Worked example

Express the ratio 40p to £2 in its simplest form.

You must ensure that the units are the same.

£2 = 200p

Then the ratio is 40 : 200 = 1 : 5 in its simplest form.

Worked example

Two lengths are in the ratio 4 : 5. If the first length is 60 cm, what is the second length?

The ratio is 4 : 5.

4 : 5 = 1 : 1.25	Dividing both sides by 4 to make an equivalent ratio, in the form 1 : n.
= 1 × 60 : 1.25 × 60	Multiplying both sides by 60 to find an equivalent ratio in the form 60 : m.
= 60 cm : 75 cm	As 1.25 × 60 = 75

So the second length is 75 cm.

≫ Proportional parts

To share an amount into proportional parts, add up the individual parts and divide the amount by this number to find the value of one part.

Worked example

Divide £50 between two sisters in the ratio 3 : 2. How much does each get?

Number of parts = 3 + 2 = 5

Value of each part = £50 ÷ 5 = £10

The sisters receive £30 (3 parts at £10 each) and £20 (2 parts at £10 each).

> **+ HINT**
>
> ▸ It is useful to check that the separate amounts add up to the original amount (i.e. £30 + £20 = £50).

? CHECK YOURSELF QUESTIONS

Q1 Express the ratio of 5 km to 600 m in its simplest form.

Q2 Express the ratio 2 : 5 in the form 1 : n.

Q3 Express the ratio $\frac{1}{3} : \frac{1}{4}$ in its simplest form.

Q4 Three children raise money for a charity. The amounts they each raise are in the ratio 2 : 3 : 7. The total amount raised is £72. How much does each child raise?

Answers are on page 187.

NUMBER

Mental methods

≫ What must I do in my head?

On the non-calculator paper, you will not be allowed to use a calculator. You will need to know **integer square numbers** from 1 × 1 up to 15 × 15, the corresponding **square roots** and the **cubes** of 2, 3, 4, 5 and 10. The mental methods described in this session will also be helpful.

≫ Multiplication

Worked example
Calculate 147 × 32.

$$
\begin{array}{ll}
147 \times 32 & = 147 \times (30 + 2) \\
& = 147 \times 30 + 147 \times 2 \\
& = 4410 + 294 \\
& = 4704
\end{array}
$$

$$
\begin{array}{r}
147 \\
\times\ 30 \\
\hline
4410
\end{array}
\qquad
\begin{array}{r}
147 \\
\times\ 2 \\
\hline
294
\end{array}
$$

It is more usual to set this multiplication out as follows.

$$
\begin{array}{rl}
147 & \\
\times\ \ 32 & \\
\hline
4410 & \text{Multiplying 147 by 30.} \\
+\ \ 294 & \text{Multiplying 147 by 2.} \\
\hline
4704 & \text{Adding.}
\end{array}
$$

MULTIPLICATION OF DECIMALS
To multiply two decimals without using a calculator:

- ignore the decimal points and multiply the numbers

- count the number of digits after the decimal point in each number and add them together, to find the number of digits after the decimal point in the answer

- place the decimal point in the answer to give the required number of digits after it.

Worked example
Find the product of 1.47 and 3.2.

147 × 32 = 4704   Ignore the decimal points and multiply the numbers, as in the previous example.

1.47 has two digits after the decimal point.

3.2 has one digit after the decimal point.

The total number of digits after the decimal points is 2 + 1 = 3, so you need 3 decimal places in the answer.

1.47 × 3.2 = 4.704

Place the decimal point so that there are three digits after the decimal point.

Worked example

Find the product of 0.000 147 and 0.0032.

$147 \times 32 = 4704$ Ignore the decimal points and multiply the numbers as in the earlier example.

The total number of digits after the decimal points = 6 + 4 = 10.
$0.000\,147 \times 0.0032 = 0.000\,000\,470\,4$

✚ *HINT*

▸ Place the decimal point so that there are 10 digits after the decimal point in the answer.

≫ Division

To divide by a two-digit number, proceed exactly as for any other division.

Worked example

Calculate $437 \div 19$.

```
      23
 19)437
     38      Take the first two digits, 43 ÷ 19 gives 2 (with a remainder of 5).
     57      Subtract 43 − 38, bring down the next digit.
     57      57 ÷ 19 gives 3 (with no remainder).
      0
```

So $437 \div 19 = 23$.

DIVISION OF DECIMALS

Use the idea of equivalent fractions to deal with division of decimals.

Worked example

Calculate $43.7 \div 1.9$.

$43.7 \div 1.9$ can be written as $\dfrac{43.7}{1.9} = \dfrac{437}{19}$

$$\dfrac{43.7}{1.9} \overset{\times 10}{=} \dfrac{437}{19} \underset{\times 10}{}$$

Work out $43.7 \div 1.9$ as $437 \div 19 = 23$ As found in the previous example.

Worked example

Work out $0.003\,08 \div 0.000\,14$.

$0.003\,08 \div 0.000\,14$ can be written as $\dfrac{0.003\,08}{0.000\,14} = \dfrac{308}{14}$ or $308 \div 14$.

```
      22
 14)308
     28      Take the first two digits, 30 ÷ 14 gives 2 (with a remainder of 2).
     28      Subtract 30 − 28, bring down the next digit.
     28      28 ÷ 14 gives 2 (with no remainder).
      0
```

So $0.003\,08 \div 0.000\,14 = 22$.

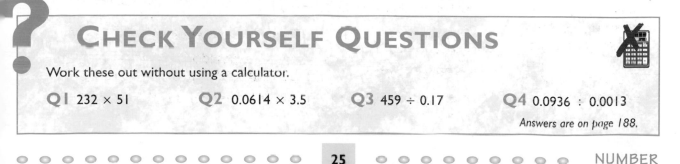

❓ CHECK YOURSELF QUESTIONS

Work these out without using a calculator.

Q1 232×51 Q2 0.0614×3.5 Q3 $459 \div 0.17$ Q4 $0.0936 \div 0.0013$

Answers are on page 188.

≫ What can my calculator do?

Make sure you are know how to use your calculator *before* the examination.

■ Read the user manual.

The following functions are to be found on most calculators, but your manual will provide further information.

■ You may need to use **INV** or **2ndF** to access some of the functions.

Key	Explanation
C	Cancel – cancels only the last number entered.
AC	All cancel – cancels all of the data entered.
x^2	Calculates the square of the number.
x^3	Calculates the cube of the number.
$\sqrt{}$	Calculates the square root of the number.
$\sqrt[3]{}$	Calculates the cube root of the number.
$1/x$ or **x^{-1}**	Calculates the reciprocal of the number.
+/−	Reverses the sign by changing positive numbers to negative numbers and negative numbers to positive numbers.
x^y	This is the power key. To enter 3^6 you key in **3** **x^y** **6**
EXP or **EE**	This is the standard-form button. To enter 3.2×10^7 you key in **3** **.** **2** **EXP** **7** The display will show 3.2 ^{07} or 3.2 ⎵07
$a^b/_c$	This is the fraction key (not all calculators have this key). To enter $\frac{3}{4}$ key in **3** **$a^b/_c$** **4** . 3⌐4 in the display means $\frac{3}{4}$. To enter $1\frac{3}{4}$ key in **1** **$a^b/_c$** **3** **$a^b/_c$** **4** . 1⌐3⌐4 in the display means $1\frac{3}{4}$.
Min or **STO**	Stores the displayed value in the memory.
MR or **RCL**	Recalls the value stored in the memory.
M+	Adds the displayed value to the number in the memory.
M−	Subtracts the displayed value from the number in the memory.
Mode	Gives the mode for calculations – refer to your user manual.
DRG	Gives the units for angles (degrees, radians or grads). Your calculator should normally be set to degrees.

CHECK YOURSELF QUESTIONS

Use your calculator to work these out.

Q1 5.1^2

Q2 $\sqrt[3]{41.3}$

Q3 2^{-5}

Q4 $(2.5 \times 10^3) \div 5$

Q5 $(2.1 \times 10^{-3}) + (4.62 \times 10^{-2})$

Q6 $\frac{4}{7} \div \frac{8}{9}$

Q7 $2\frac{1}{5} \times \frac{5}{9}$

Answers are on page 188.

Compound measures

≫ What are compound measures?

Compound measures involve more than one unit such as:

- speed (distance and time) or
- density (mass and volume).

≫ Speed

The formula for speed is:

- speed = $\dfrac{\text{distance}}{\text{time}}$ (expressed in units such as miles per hour)

- The formula for speed can be rearranged to give:
  - distance = speed × time
  - time = $\dfrac{\text{distance}}{\text{speed}}$

Worked example

A taxi travels 16 miles in 20 minutes.
What is the speed in miles per hour?

Distance = 16 miles

Time = 20 minutes = $\frac{20}{60}$ = $\frac{1}{3}$ hour

Using the formula speed = $\dfrac{\text{distance}}{\text{time}}$:

speed = $\dfrac{16}{\frac{1}{3}}$

       = 48 miles per hour (mph)

The speed is 48 mph.

Worked example

A cyclist travels 3.6 km at an average speed of 8 kilometres per hour.
How long does the journey take?

Using the formula time = $\dfrac{\text{distance}}{\text{speed}}$:

time = $\frac{3.6}{8}$ = 0.45 hours

0.45 hours = 0.45 × 60 minutes
           = 27 minutes

The journey takes 27 minutes.

+ HINT

- Speed is measured in miles per hour, so express the distance in miles and the time in hours.

+ HINT

- Remember that 0.45 hours is not 45 minutes as there are 60 minutes in one hour.
- To convert hours to minutes, multiply by 60.

≫ Density

The formula for density is:

- density = $\dfrac{\text{mass}}{\text{volume}}$ (expressed in units such as grams per cm^3)

- The formula for density can be rearranged to give:
  - mass = density × volume
  - volume = $\dfrac{\text{mass}}{\text{density}}$

+ HINT

- Density is measured in g/cm^3, so express the mass in grams and the volume in cm^3.

Worked example

A piece of lead weighing 170 g has a volume of 15 cm^3.
Give an estimate for the density of lead.

Using the formula density = $\dfrac{\text{mass}}{\text{volume}}$:

density = $\dfrac{170}{15}$ = 11.3 g/cm^3 (3 s.f.)

The density of lead is about 11.3 g/cm^3 (3 s.f.).

Worked example

A sheet of metal measures 10 m by 6 m by 0.5 mm and has a density of 8.9 g/cm^3.
What is the mass of the sheet of metal, in kilograms?

Volume = 1000 × 600 × 0.05 = 30 000 cm^3 Converting all lengths to cm.

Using the formula mass = density × volume:

mass = 8.9 × 30 000 g = 267 000 g = 267 kg   Converting to kg where 1000 g = 1 kg.

The mass of the sheet of metal is 267 kg.

? CHECK YOURSELF QUESTIONS

Q1 A train travels 160 metres in 10 seconds. What is its speed in **a)** m/s **b)** km/h?

Q2 A cyclist travels for 45 minutes at a speed of 14 miles per hour. What distance does the cyclist travel?

Q3 A silver cube of length 12 mm has a density of 10.5 g/cm^3. What is the mass of the cube?

Answers are on page 188.

≫ What is the difference?

With **simple interest, the interest paid is not reinvested,** whereas **with compound interest, the amount of interest paid is reinvested and earns interest** itself.

The formula for simple interest is:

■ $A = P + \dfrac{PRT}{100}$

and the formula for compound interest is:

■ $A = P \times (1 + \dfrac{R}{100})^T$

where: A = total amount
P = principal or original investment
R = rate (% per annum)
T = time (in years)

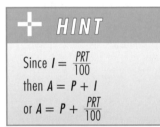

+ HINT

Since $I = \dfrac{PRT}{100}$
then $A = P + I$
or $A = P + \dfrac{PRT}{100}$

≫ Simple interest

Simple interest is calculated on a fixed principal, and can be calculated by the simple formula:

■ $I = \dfrac{PRT}{100}$

where: I = interest
P = principal
R = rate (% per annum or per year)
T = time (in years)

Worked example

If £2000 is invested for 2 years at 6% per annum, calculate the simple interest and the total amount.

Using the formula $I = \dfrac{PRT}{100}$

where: P = principal = £2000
R = rate = 6%
T = time = 2 years

$I = \dfrac{2000 \times 6 \times 2}{100}$

  = £240

$A = P + I$

  = £2000 + £240

A = £2240

The simple interest is £240 and the total amount is £2240.

≫ Compound interest

In compound interest, the principal changes every year, as the previous year's interest is added into it.

Worked example

£1000 is invested at 6.2% p.a. compound interest. Find the amount after 3 years.

The compound interest can be worked out by repeatedly applying the simple interest formula:

$$A = P + \frac{PRT}{100}$$

where: P = principal = £1000
R = rate = 6.2%
T = time = 1 year for each year

Year 1 $A = P + \frac{PRT}{100} = 1000 + \frac{1000 \times 6.2 \times 1}{100} = £1062$

Year 2 $A = P + \frac{PRT}{100}$ After 1 year, P = £1062.

$= 1062 + \frac{1062 \times 6.2 \times 1}{100} = £1127.844$

Year 3 $A = P + \frac{PRT}{100}$ After 2 years, P = £1127.844.

$= 1127.844 + \frac{1127.844 \times 6.2 \times 1}{100} = £1197.770\ 328$

$= £1197.77$ (to the nearest penny)

+ HINT

▸ Do not round off until the final answer.

Alternatively, using the compound interest formula:

$$A = P \times (1 + \frac{R}{100})^T$$

where: P = principal = £1000
R = rate = 6.2%
T = time = 3 years

$A = 1000 (1 + \frac{6.2}{100})^3 = £1197.770\ 328$ (as before)

$A = £1197.77$ (to the nearest penny)

? CHECK YOURSELF QUESTIONS

Q1 A sum of £5000 is invested at 7% p.a. simple interest. How long will it be before the amount equals £5875?

Q2 A sum of £250 is invested for 4 years. The simple interest paid is £52.50. What is the percentage rate per annum?

Q3 A sum of £2000 is invested for 2 years at 6% per annum. Calculate the total amount and the compound interest.

Answers are on page 189.

Rational and irrational numbers

≫ What are rational and irrational numbers?

A rational number can be expressed in the form $\frac{p}{q}$ where p and q are integers.

- **Rational numbers** include $\frac{1}{5}$, $0.\dot{3}$, 7, $\sqrt{9}$, $\sqrt[3]{64}$.
- **Irrational numbers** include $\sqrt{2}$, $\sqrt{3}$, $\sqrt[3]{20}$, π, π^2.

Irrational numbers involving square roots are also called **surds**. Surds can be multiplied and divided.

+ **HINTS**

▸ $\sqrt{a} \times \sqrt{b} = \sqrt{a \times b}$

▸ $\frac{\sqrt{a}}{\sqrt{b}} = \sqrt{\frac{a}{b}}$

Worked example

Work these out.

a) $\sqrt{3} \times \sqrt{3}$ **b)** $\sqrt{2} \times \sqrt{8}$ **c)** $\frac{\sqrt{48}}{\sqrt{12}}$ **d)** $\frac{\sqrt{30}}{\sqrt{6}}$

a) $\sqrt{3} \times \sqrt{3} = \sqrt{3 \times 3} = \sqrt{9} = 3$ **b)** $\sqrt{2} \times \sqrt{8} = \sqrt{2 \times 8} = \sqrt{16} = 4$

c) $\frac{\sqrt{48}}{\sqrt{12}} = \sqrt{\frac{48}{12}} = \sqrt{4} = 2$ **d)** $\frac{\sqrt{30}}{\sqrt{6}} = \sqrt{\frac{30}{6}} = \sqrt{5}$

Worked example

Simplify the following.

a) $\sqrt{72}$ **b)** $\sqrt{5} + \sqrt{45}$ **c)** $\frac{1}{\sqrt{5}}$ **d)** $(3 - \sqrt{2})^2$

a) $\sqrt{72} = \sqrt{36 \times 2} = \sqrt{36} \times \sqrt{2} = 6 \times \sqrt{2}$
$= 6\sqrt{2}$ As $6 \times \sqrt{2}$ is usually written $6\sqrt{2}$.

b) $\sqrt{5} + \sqrt{45} = \sqrt{5} + \sqrt{9 \times 5} = \sqrt{5} + (\sqrt{9} \times \sqrt{5})$
$= \sqrt{5} + (3 \times \sqrt{5}) = \sqrt{5} + 3\sqrt{5} = 4\sqrt{5}$

c) $\frac{1}{\sqrt{5}} = \frac{1}{\sqrt{5}} \times \frac{\sqrt{5}}{\sqrt{5}} = \frac{\sqrt{5}}{5}$ As $(\sqrt{5})^2$ is 5.

d) $(3 - \sqrt{2})^2 = 3 \times 3 - 3\sqrt{2} - 3\sqrt{2} + \sqrt{2}\sqrt{2}$
$= 9 - 6\sqrt{2} + 2$
$= 11 - 6\sqrt{2}$

? CHECK YOURSELF QUESTIONS

Q1 Which of these are rational numbers?

$3^{\frac{1}{2}}$ $(\sqrt{3})^2$ π^{-2} $\sqrt{5\frac{1}{4}}$ $\sqrt{6\frac{1}{4}}$

Write each of the rational numbers in the form $\frac{p}{q}$ where p and q are integers.

Q2 Simplify the following expressions, leaving your answers in surd form.

a) $\sqrt{5} \times \sqrt{15}$ **b)** $\sqrt{5} + \sqrt{20}$ **c)** $\frac{1}{\sqrt{7}}$

Q3 Simplify $(4 + \sqrt{3})(4 - \sqrt{3})$. (See *Expanding brackets* on page 43.)

Answers are on page 189.

Recurring decimals

HIGHER

≫ What are recurring decimals?

In recurring decimals, part of the decimal fraction is repeated indefinitely.

Recurring decimals are all rational numbers as they can be expressed as fractions.

- 0.166 666 666 ...   written 0.1$\dot{6}$ = $\frac{1}{6}$

- 0.142 857 142 857 ...   written 0.$\dot{1}$42 85$\dot{7}$ = $\frac{1}{7}$

- 0.777 777 777 ...   written 0.$\dot{7}$ = $\frac{7}{9}$

- 0.272 727 27 ...   written 0.$\dot{2}\dot{7}$ = $\frac{3}{11}$

CONVERTING RECURRING DECIMALS

Worked example
Change 0.$\dot{8}$ to a fraction.

Notice that	$10 \times 0.\dot{8} = 8.888\,888\,8...$	Multiplying both sides by 10.
and	$1 \times 0.\dot{8} = 0.888\,888\,8...$	
Subtracting:	$9 \times 0.\dot{8} = 8$	$8.888\,888\,8... - 0.888\,888\,8...$
and	$0.\dot{8} = \frac{8}{9}$	Dividing both sides by 9.
So	$0.\dot{8} = \frac{8}{9}$	

+ HINT

▶ Now check this by putting $\frac{8}{9}$ into your calculator.

Worked example
Convert 14.2$\dot{3}$ to a mixed number.

A mixed number consists of a whole number part and a fractional part.
In this question you can split the number up and deal with the recurring decimal or else proceed as shown previously.

Notice that	$100 \times 14.2\dot{3} = 1423.232\,323\,...$	Multiplying both sides by 100.
and	$1 \times 14.2\dot{3} = 14.232\,323\,...$	
Subtracting:	$99 \times 14.2\dot{3} = 1409$	$1423.232\,323\,... - 14.232\,323\,...$
and	$14.2\dot{3} = \frac{1409}{99}$	Dividing both sides by 99.
	$= 14\frac{23}{99}$	Converting back to a mixed number.
So	$14.2\dot{3} = 14\frac{23}{99}$	

? CHECK YOURSELF QUESTIONS

Q1 Write a fraction that is equivalent to the recurring decimal 0.25$\dot{3}$.

Q2 Change 0.8$\dot{3}\dot{5}$ to a fraction.

Answers are on page 189.

Fractional indices

≫ Rules for fractional indices

For any real numbers a, m and n (See Revision session 5, *Positive, negative and zero indices*):

- $a^m \times a^n = a^{m+n}$
- $a^1 = a$
- $a^m \div a^n = a^{m-n}$
- $a^0 = 1$
- $a^{-m} = \dfrac{1}{a^m}$

Using the rules above, you can also see that:

- $a^{\frac{1}{2}} \times a^{\frac{1}{2}} = a^{\frac{1}{2} + \frac{1}{2}} = a^1 = a$

  so any number raised to the power $\frac{1}{2}$ means $\sqrt[2]{}$ or $\sqrt{}$ $a^{\frac{1}{2}} = \sqrt{a}$

- $a^{\frac{1}{3}} \times a^{\frac{1}{3}} \times a^{\frac{1}{3}} = a^{\frac{1}{3} + \frac{1}{3} + \frac{1}{3}} = a^1 = a$

  so any number raised to the power $\frac{1}{3}$ means $\sqrt[3]{}$ $a^{\frac{1}{3}} = \sqrt[3]{a}$

Similarly:

- $a^{\frac{1}{n}} = \sqrt[n]{a}$

CALCULATIONS WITH FRACTIONAL INDICES

Worked example

Evaluate the following.

a) $49^{\frac{1}{2}}$

$$49^{\frac{1}{2}} = \sqrt{49}$$
$$= \pm 7$$

Remember to give both roots, even though your calculator will only give the positive root.

b) $3125^{\frac{1}{5}}$

$$3125^{\frac{1}{5}} = \sqrt[5]{3125}$$
$$= 5$$

Worked example

Evaluate $216^{\frac{2}{3}}$.

$$216^{\frac{2}{3}} = (216^{\frac{1}{3}})^2$$
$$= 6^2 \text{ as } 216^{\frac{1}{3}} = 6$$
$$= 36$$

Alternatively, you can use: $216^{\frac{2}{3}} = (216^2)^{\frac{1}{3}} = 46656^{\frac{1}{3}} = 36$ although this is rather longwinded.

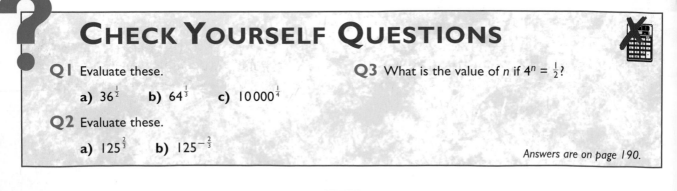

CHECK YOURSELF QUESTIONS

Q1 Evaluate these.

    a) $36^{\frac{1}{2}}$ **b)** $64^{\frac{1}{3}}$ **c)** $10\,000^{\frac{1}{4}}$

Q2 Evaluate these.

    a) $125^{\frac{2}{3}}$ **b)** $125^{-\frac{2}{3}}$

Q3 What is the value of n if $4^n = \frac{1}{2}$?

Answers are on page 190.

REVISION SESSION 19
HIGHER
Direct and inverse proportion

≫ How do you write proportions in maths?

With **direct proportion** as one variable increases, the other increases and as one decreases, the other one decreases.

With **inverse proportion** as one variable increases, the other decreases and as one decreases, the other one increases.

- If y is proportional to x then $y \propto x$ or $y = kx$.
- If y is inversely proportional to x then $y \propto \frac{1}{x}$ or $y = \frac{k}{x}$.

The value of k is a constant and k is called the **constant of proportionality**.

≫ Working with proportion

Worked example

Rewrite these statements, using the $\propto$ sign and the constant of proportionality.

a) y varies directly as x cubed.

b) s is proportional to the square root of t.

c) v varies inversely as the square of w.

d) p is inversely proportional to the cube root of q.

a) $y \propto x^3$ or $y = kx^3$ **b)** $s \propto \sqrt{t}$ or $s = k\sqrt{t}$

c) $v \propto \frac{1}{w^2}$ or $v = \frac{k}{w^2}$ **d)** $p \propto \frac{1}{\sqrt[3]{q}}$ or $p = \frac{k}{\sqrt[3]{q}}$

Worked example

Given that a varies directly as the cube of b and $a = 4$ when $b = 2$, find the value of k and the value of a when $b = 3$.

If a varies directly as the cube of b then $a \propto b^3$ or $a = kb^3$.

Since we know that $a = 4$ when $b = 2$, then: $4 = k \times 2^3$
$$4 = k \times 8$$
$$\text{so } k = \tfrac{1}{2}$$

The equation is $a = \frac{1}{2}b^3$.

When $b = 3$ then $a = \frac{1}{2} \times 3^3 = \frac{1}{2} \times 27 = 13.5$.

CHECK YOURSELF QUESTIONS

Q1 Given that T is proportional to the positive square root of W and $T = 36$ when $W = 16$:

    a) calculate T when W is 100

    b) calculate W when T is 18.

Q2 If V varies inversely as the cube of Y and $V = \frac{3}{8}$ when $Y = 2$, find the value of V when:

    a) $Y = 3$ **b)** $Y = 10$.

Answers are on page 190.

≫ What are upper and lower bounds?

If a weight is given as 10 grams to the nearest gram, then the actual weight will lie in the interval 9.5 grams to 10.499 999... grams as all values in this interval will be rounded to 10 grams to the nearest gram. The weight 10.499 999... grams is usually written as 10.5 g although it is accepted that 10.5 g would be rounded to 11 g (to the nearest gram).

- The value 9.5 g is called the **lower bound** as it is the lowest value which would be rounded to 10 g, while 10.5 g is called the **upper bound**.

≫ Working with bounds

Worked example

A rectangle measures 10 cm by 6 cm, where each measurement is given to the nearest cm. Write down an interval approximation for the area of a rectangle.

The lower bound (minimum area) = $9.5 \times 5.5 = 52.25 \, cm^2$

The upper bound (maximum area) = $10.5 \times 6.5 = 68.25 \, cm^2$

The interval approximation = $52.25 \, cm^2$ to $68.25 \, cm^2$

Worked example

To the nearest whole number, the value of p is 215 and the value of q is 5. Calculate the maximum and minimum values of the following expressions.

a) $p + q$ **b)** $p - q$
c) $p \times q$ **d)** $p \div q$

| p_{min} = 214.5 p_{max} = 215.5 | q_{min} = 4.5 q_{max} = 5.5 |

a) For $p + q$
maximum = 215.5 + 5.5 = 221
minimum = 214.5 + 4.5 = 219

b) For $p - q$
maximum = 215.5 − 4.5 = 211
minimum = 214.5 − 5.5 = 209

c) For $p \times q$
maximum = 215.5 × 5.5 = 1185.25
minimum = 214.5 × 4.5 = 965.25

d) For $p \div q$
maximum = 215.5 ÷ 4.5 = 47.888 ...
minimum = 214.5 ÷ 5.5 = 39

+ HINTS

- For the maximum value of $p - q$, work out $p_{max} - q_{min}$.
- For the minimum value of $p - q$, work out $p_{min} - q_{max}$.
- For the maximum value of $p \div q$, work out $p_{max} \div q_{min}$.
- For the minimum value of $p \div q$, work out $p_{min} \div q_{max}$.

? CHECK YOURSELF QUESTIONS

Q1 The length of a square is 5.2 cm, correct to two significant figures. Find the maximum and minimum values for its area.

Q2 A Ford Scorpio costs £18 700 and a Vauxhall Calibra costs £17 300, both prices being given to the nearest £100. What is the least possible difference in price between the two cars?

Answers are on page 190.

UNIT 2: ALGEBRA

Patterns and sequences

≫ What is a sequence?

A **sequence** is a set of numbers that follow a particular rule. The word **term** is often used to describe the numbers in the sequence. In the following sequence, the first term is 7, the second term is 9, the third term is 11, and so on.

   7, 9, 11, 13, 15, 17, ...

The **general form** of the terms is called the **nth term**, and this gives the value of any term in the sequence. If you are given a sequence in which the nth term is $2n + 5$ then:

	first term (where $n = 1$) gives	$2 \times 1 + 5 = 7$
	second term (where $n = 2$) gives	$2 \times 2 + 5 = 9$
	third term (where $n = 3$) gives	$2 \times 3 + 5 = 11$
Similarly:	50th term (where $n = 50$) gives	$2 \times 50 + 5 = 105$
and	1000th term (where $n = 1000$) gives $2 \times 1000 + 5 = 2005$	

≫ Sequence rules

Most number sequences involve adding/subtracting or multiplying/dividing in the rule for finding one term from the one before it. Once you have found the term-to-term rule, you can use it to find subsequent terms.

Worked example
Find the sequence rule and find the next three terms of these sequences.

a) 1, 6, 11, 16, 21, ...   **b)** 243, 81, 27, 9, 3, ...
c) 1, 4, 13, 40, 121, ...

a)   1   6   11   16   21   ...
              +5   +5   +5   +5   ...

The rule for moving from term to term is + 5.
The next three terms are 26, 31 and 36.

b)   243   81   27   9   3   ...
              ÷ 3   ÷ 3   ÷ 3   ÷ 3   ...

The rule for moving from term to term is ÷ 3.
The next three terms are 1, $\frac{1}{3}$ and $\frac{1}{9}$.

c)   1   4   13   40   121   ...
              × 3 + 1   × 3 + 1   × 3 + 1   × 3 + 1   ...

The rule for moving from term to term is × 3 + 1.
The next three terms are 364, 1093 and 3280.

FINDING TERMS

You can use the term-to-term method shown above for finding subsequent terms of a sequence but it would be rather time-consuming if you needed to find the 100th term or the 500th term etc.

The following method gives you the nth term of a linear sequence.

Worked example

Find the nth term of the following linear sequences.

a) 3, 7, 11, 15, ...

b) 12, 9, 6, 3, ...

a)

$$3 \quad\underset{+4}{\overset{}{\curvearrowright}}\quad 7 \quad\underset{+4}{\overset{}{\curvearrowright}}\quad 11 \quad\underset{+4}{\overset{}{\curvearrowright}}\quad 15 \quad\underset{+4}{\overset{}{\curvearrowright}}\quad \cdots \quad \cdots$$

The difference between terms is 4, so use 4 as the multiplier.

You can construct a table as follows.

Term number		1	2	3	4
Sequence		3	7	11	15
Try 4 × (term number)		4	8	12	16
Difference		$^-1$	$^-1$	$^-1$	$^-1$
The difference is always equal to -1.					
So try 4 × (term number) $-$ 1.		3	7	11	15

So the nth term is $4 \times n - 1$ or $4n - 1$.

b)

$$12 \quad\underset{-3}{\overset{}{\curvearrowright}}\quad 9 \quad\underset{-3}{\overset{}{\curvearrowright}}\quad 6 \quad\underset{-3}{\overset{}{\curvearrowright}}\quad 3 \quad\underset{-3}{\overset{}{\curvearrowright}}\quad \cdots \quad \cdots$$

The difference between terms is $^-3$, so use $^-3$ as the multiplier.

You can construct a table as follows.

Term number		1	2	3	4
Sequence		12	9	6	3
Try ($^-3$) × (term number)		$^-3$	$^-6$	$^-9$	$^-12$
Difference		15	15	15	15
The difference is always equal to 15.					
So try ($^-3$) × (term number) + 15.		12	9	6	3

So the nth term is $^-3 \times n + 15$ or $^-3n + 15$ or $15 - 3n$.

≫ Special sequences

You should be able to recognise the following special sequences of numbers.

1, 4, 9, 16, 25, ...	square numbers
1, 3, 6, 10, 15, ...	triangle numbers
1, 8, 27, 64, 125, ...	cube numbers
2, 3, 5, 7, 11, 13, 17, ...	prime numbers

Another sequence that you should know is the **Fibonacci sequence**. Each term after the second is found by adding the two previous terms.

1, 1, 2, 3, 5, 8, 13, 21, ...

So 3rd term = 1st term + 2nd term
4th term = 2nd term + 3rd term   etc.

HINT

means 'Do not use a calculator.'

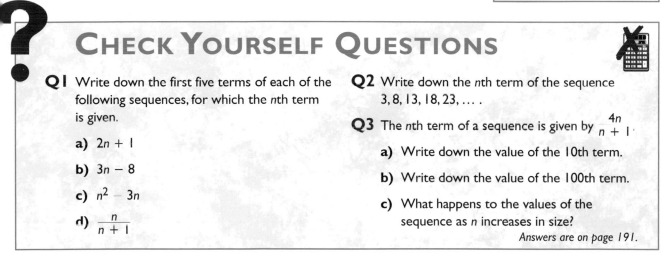

? CHECK YOURSELF QUESTIONS

Q1 Write down the first five terms of each of the following sequences, for which the *n*th term is given.

a) $2n + 1$

b) $3n - 8$

c) $n^2 - 3n$

d) $\dfrac{n}{n + 1}$

Q2 Write down the *n*th term of the sequence 3, 8, 13, 18, 23, ...

Q3 The *n*th term of a sequence is given by $\dfrac{4n}{n + 1}$.

a) Write down the value of the 10th term.

b) Write down the value of the 100th term.

c) What happens to the values of the sequence as *n* increases in size?

Answers are on page 191.

Substitution

≫ What is substitution?

Substitution means replacing the letters in a formula or expression by the given numbers.

Worked example

Find the value of:

a) $3x + 4y - 5z$ where $x = 3, y = 7$ and $z = {}^-4$

b) $3ab + ac^2 - \dfrac{b}{c}$ where $a = 2, b = {}^-10$ and $c = 5$.

a) Substituting $x = 3, y = 7$ and $z = {}^-4$ in $3x + 4y - 5z$:

$$3x + 4y - 5z = 3 \times 3 + 4 \times 7 - 5 \times {}^-4$$
$$= 9 + 28 - {}^-20 \qquad \text{Remember that } -(-) = +.$$
$$= 9 + 28 + 20$$
$$= 57$$

b) Substituting $a = 2, b = {}^-10$ and $c = 5$ in $3ab + ac^2 - \dfrac{b}{c}$:

$$3ab + ac^2 - \frac{b}{c} = 3 \times 2 \times {}^-10 + 2 \times 5 \times 5 - \frac{{}^-10}{5}$$
$$= {}^-60 + 50 - {}^-2 \quad \text{Remember that } -(-) = +.$$
$$= {}^-60 + 50 + 2$$
$$= {}^-8$$

> **HINT**

▸ Remember: $3x$ means $3 \times x$, $4y$ means $4 \times y$ and $5z$ means $5 \times z$.

> **HINT**

▸ Remember:
$ac^2 = a \times c \times c$
so only the c is squared.

? CHECK YOURSELF QUESTIONS

Q1 The surface area, A, of a closed cone is given by $A = \pi rl + \pi r^2$ where r is the radius of the base and l is the slant height.

Find the surface area of a cone with radius 3 cm and slant height 8 cm, giving your answer in terms of π.

Q2 Given that $\dfrac{1}{u} = \dfrac{1}{f} - \dfrac{1}{v}$, find the value of u when:

a) $f = 4$ and $v = 8$

b) $f = 3$ and $v = 5$.

Answers are on page 191.

■ Simplifying expressions ■

≫ What is an expression?

An **algebraic expression** is a collection of algebraic terms along with their + and − signs.

≫ Like terms

Like terms are numerical multiples of the same algebraic quantity. For example, $3x$, $-5x$, $\frac{1}{2}x$ and $0.55x$ are all like terms because they are all multiples of the same algebraic quantity, x.

Worked example

Collect together the like terms from the following list.

x	$3x$	y
^-6x	xy^2	$3ab$
^-3y	xy	^-4xy
$\frac{1}{3}y$	x^2y	$7ab$
^-2ba	yx^2	$-\frac{3}{4}xy^2$

Like terms are numerical multiples of the same algebraic quantity.

x, $3x$, ^-6x	are all terms in x
y, ^-3y, $\frac{1}{3}y$	are all terms in y
xy^2, $-\frac{3}{4}xy^2$	are all terms in xy^2
$3ab$, $7ab$, ^-2ba	are all terms in ab where $^-2ba = {}^-2ab$
xy, ^-4xy	are all terms in xy
x^2y, yx^2	are all terms in x^2y where $yx^2 = x^2y$

In algebraic terms involving more than one letter, it is useful to write them in alphabetical order, so that $3ba$ is written $3ab$ and $14zxy$ is written $14xyz$ etc.

≫ Adding, subtracting, multiplying and dividing

You can add or subtract like terms. The process of adding and subtracting like terms in an expression or equation is called **simplifying**.

Worked example

Simplify the following expression.
$3x + 2y - 7z + 4x - 3y$

$3x + 2y - 7z + 4x - 3y$	
$= 3x + 4x + 2y - 3y - 7z$	Putting like terms together, *along with the signs in front of them.*
$= 7x - 1y - 7z$	Adding and subtracting like terms.
$= 7x - y - 7z$	Rewriting $1y$ as y.

Worked example

Simplify the following expression.

$4p + 3pq - 3p + 8qp - 2r$

$4p + 3pq - 3p + 8qp - 2r$	
$= 4p - 3p + 3pq + 8qp - 2r$	Putting like terms together, *along with the signs in front of them.*
$= 4p - 3p + 3pq + 8pq - 2r$	Writing $8qp$ as $8pq$ since $p \times q$ is the same as $q \times p$.
$= p + 11pq - 2r$	

You can also simplify terms or expressions by multiplying or dividing. When you are dividing terms, you can simplify them by cancelling.

Worked example

Simplify the following expressions.

a) $3f \times 4g$

b) $8pq^2 \div 2pq$

a)	$3f \times 4g = 3 \times f \times 4 \times g$	
	$= 12 \times f \times g$	Multiplying $3 \times 4 = 12$.
	$= 12fg$	Rewriting without the $\times$ signs.
b)	$8pq^2 \div 2pq = \dfrac{8 \times p \times q \times q}{2 \times p \times q}$	Writing as a fraction with $q^2 = q \times q$.
	$= \dfrac{\overset{4}{\cancel{8}} \times \overset{1}{\cancel{p}} \times q \times \overset{1}{\cancel{q}}}{\underset{1}{\cancel{2}} \times \underset{1}{\cancel{p}} \times \underset{1}{\cancel{q}}}$	Cancelling.
	$= 4q$	Rewriting without the $\times$ signs.

CHECK YOURSELF QUESTIONS

Q1 Simplify the following expressions.

a) $3a + 6b - 2a - 5c$

b) $5x + 7y - 3xy + 2x + 2yx$

c) $x^3 + 3x^2 - 4x - 9x^2 + 7x - 2$

d) $3x \times 4y \times 2z$

e) $5a \times 2a^2$

f) $(4abc)^2 \times a^2b$

g) $8mn^3 \div 4n$

Answers are on page 191.

REVISION SESSION 4 — Expanding and factorising

>> What do the words mean?

Brackets are used to group algebraic terms. The process of removing brackets from an expression (or an equation) is called **expanding** and the process of rewriting an expression (or an equation) so that it includes terms in brackets is called **factorising**.

>> Expanding brackets

When expanding brackets in an expression, you must multiply all the terms inside the brackets by the term just before the bracket.

+ **HINT**

▸ You must take care, when expanding brackets, to multiply every term in the brackets by the term outside the brackets.

Worked example

Expand the following expression.

$3(5a - 2b)$

$3(5a - 2b) = 3 \times 5a + 3 \times {}^-2b$
$= 15a - 6b$

Worked example

Expand the following expression.

${}^-7(a + 2b - 3c)$

${}^-7(a + 2b - 3c) = {}^-7 \times a - 7 \times 2b - 7 \times {}^-3c$
$= {}^-7a - 14b + 21c$ Remembering that ${}^-7 \times {}^-3c = {}^+21c$.

Worked example

Expand and simplify the following expression.

$5(p + 2q) + 2(p - 5q)$

$5(p + 2q) + 2(p - 5q) = 5 \times p + 5 \times 2q + 2 \times p + 2 \times {}^-5q$
$= 5p + 10q + 2p - 10q$
$= 7p + 0q$
$= 7p$

Worked example

Expand and simplify the following expression.

$4a(b - c) - 3b(a - c)$

$4a(b - c) - 3b(a - c) = 4a \times b + 4a \times {}^-c - 3b \times a - 3b \times {}^-c$
  The term outside the second bracket is ${}^-3b$.
$= 4ab - 4ac - 3ab + 3bc$
  Writing $3ba$ as $3ab$ and remembering
  that ${}^-3b \times {}^-c = {}^+3bc$.
$= 1ab - 4ac + 3bc$
$= ab - 4ac + 3bc$ As $1ab$ is usually written ab.

≫ Factorising expressions into brackets

To factorise an expression, you need to look for terms that have **common factors**. Then you rewrite the expression with the factors outside brackets. Remember that common factors of two (or more) terms are factors that appear in both (or all) of the terms.

Worked example

Factorise $5a - 15$.

To factorise the expression, write it as a term inside brackets, with common factors taken outside the brackets.

$5a - 15 = 5(a - 3)$ 5 is a common factor, as
$5a = 5 \times a$ and $15 = 5 \times 3$.

Worked example

Factorise $9xy + 33yz$.

$9xy + 33yz = 3y(3x + 11z)$ $3y$ is a common factor as
$9xy = 3y \times 3x$ and $33yz = 3y \times 11z$.

Worked example

Factorise $pq^3 - p^2q$.

$pq^3 - p^2q = pq(q^2 - p)$ pq is a common factor as
$pq^3 = pq \times q^2$ and $p^2q = pq \times q$.

≫ Binomial expressions

A binomial expression consists of two terms such as $(a + b)$ or $(5x - 2z)$.

To expand the product of two binomial expressions, you must multiply each term in the first expression by each term in the second expression.

$$(a + b)(c + d) = a \times (c + d) + b \times (c + d)$$
$$= a \times c + a \times d + b \times c + b \times d$$
$$= ac + ad + bc + bd$$

		Product
F = First	$(a + b)(c + d)$	$a \times c$
O = Outsides	$(a + b)(c + d)$	$a \times d$
I = Insides	$(a + b)(c + d)$	$b \times c$
L = Last	$(a + b)(c + d)$	$b \times d$

Worked example

Expand $(x + 3)(x + 5)$.

$$(x + 3)(x + 5) = x \times x + x \times 5 + 3 \times x + 3 \times 5$$
$$= x^2 + 5x + 3x + 15$$
$$= x^2 + 8x + 15$$

Worked example

Expand $(3x - 1)(4x - 5)$.

$$(3x - 1)(4x - 5) = 3x \times 4x + 3x \times {}^-5 + {}^-1 \times 4x + {}^-1 \times {}^-5$$
$$= 12x^2 - 15x - 4x + 5 \text{ Remembering that } {}^-1 \times {}^-5 = {}^+5.$$
$$= 12x^2 - 19x + 5$$

You can use the reverse process to write a quadratic as a product of brackets, as shown in the following worked example.

Worked example

Factorise $x^2 + 5x + 6$.

You know that $x^2 + 5x + 6 = (x \quad)(x \quad)$ Since $x \times x = x^2$ as required.

Look for pairs of numbers that multiply together to give ${}^+6$.

Possibilities include:

${}^+1 \times {}^+6$ $(x + 1)(x + 6) = x^2 + 6x + 1x + 6$
$\qquad\qquad\qquad\qquad\quad = x^2 + 7x + 6$

${}^-1 \times {}^-6$ $(x - 1)(x - 6) = x^2 - 6x - 1x + 6$
$\qquad\qquad\qquad\qquad\quad = x^2 - 7x + 6$

${}^+2 \times {}^+3$ $(x + 2)(x + 3) = x^2 + 2x + 3x + 6$
$\qquad\qquad\qquad\qquad\quad = x^2 + 5x + 6$

${}^-2 \times {}^-3$ $(x \quad 2)(x - 3) = x^2 - 2x - 3x + 6$
$\qquad\qquad\qquad\qquad\quad = x^2 - 5x - 6$

and the correct solution is:

$$(x + 2)(x + 3) = x^2 + 2x + 3x + 6$$
$$= x^2 + 5x + 6.$$

An alternative method is to look at pairs of numbers that multiply together to give ${}^+6$ (i.e. with product ${}^+6$) and add to give ${}^+5$ (i.e. with sum ${}^+5$).

Numbers	Product	Sum	
${}^+1$ and ${}^+6$	${}^+6$	${}^+7$	✗
${}^-1$ and ${}^-6$	${}^+6$	${}^-7$	✗
${}^+2$ and ${}^+3$	${}^+6$	${}^+5$	✔
${}^-2$ and ${}^-3$	${}^+6$	${}^-5$	✗

From this, you can quickly see the numbers you need.

Again, the correct solution is $(x + 2)(x + 3)$.

Worked example

Factorise $x^2 + x - 12$.

You know that $x^2 + x - 12 = (x \quad)(x \quad)$ Since $x \times x = x^2$ as required.

You now need to look at pairs of numbers which multiply together to give $^-12$.

Possibilities include:

$$^-1 \times {}^+12 \quad (x - 1)(x + 12) = x^2 - 1x + 12x - 12 \quad = x^2 + 11x - 12$$

$$^+1 \times {}^-12 \quad (x + 1)(x - 12) = x^2 + 1x - 12x - 12 = x^2 - 11x - 12$$

$$^-2 \times {}^+6 \quad (x - 2)(x + 6) = x^2 - 2x + 6x - 12 \quad = x^2 + 4x - 12$$

$$^+2 \times {}^-6 \quad (x + 2)(x - 6) = x^2 + 2x - 6x \quad 12 \quad = x^? - 4x - 12$$

$$^-3 \times {}^+4 \quad (x - 3)(x + 4) = x^2 - 3x + 4x - 12 \quad = x^2 + x - 12$$

$$^+3 \times {}^-4 \quad (x + 3)(x - 4) = x^2 + 3x - 4x - 12 \quad = x^2 - x - 12$$

So the correct solution is:

$$(x - 3)(x + 4) = x^2 - 3x + 4x - 12 = x^2 + x - 12.$$

Again, using the alternative method, look for numbers with product $^-12$ and sum $^+1$.

Numbers	Product	Sum	
$^-1$ and $^+12$	$^-12$	$^+11$	✗
$^+1$ and $^-12$	$^-12$	$^-11$	✗
$^-2$ and $^+6$	$^-12$	$^+4$	✗
$^+2$ and $^-6$	$^-12$	$^-4$	✗
$^-3$ and $^+4$	$^-12$	$^+1$	✔
$^+3$ and $^-4$	$^-12$	$^-1$	✗

The correct solution is $(x - 3)(x + 4)$.

+ HINT

▶ $(x - 3)(x + 4)$ can be written as $(x + 4)(x - 3)$, which also gives $x^2 + x - 12$.

CHECK YOURSELF QUESTIONS

Q1 Expand and simplify these expressions.

a) $5x - (3y + 4x)$

b) $5(a + b - 2c) - 2(a - 2b + 3c)$

c) $a(2a + b) - 2b(a - b^2)$

Q2 Factorise these expressions completely.

a) $4x^2 - 6x$

b) $2lw + 2wh + 2hl$

c) $5x^2y - 10xy^2$

Q3 Expand and simplify these expressions.

a) $(x + 1)(x + 4)$

b) $(3y - 5)(2y - 7)$

c) $(3x + 1)^2$

Q4 Factorise these expressions completely.

a) $x^2 + 6x + 8$

b) $x^2 - x - 2$

c) $x^2 - 7x - 18$

Answers are on page 192.

REVISION SESSION 5 — Solving equations

≫ What is an equation?

An **algebraic equation** is made up of two algebraic expressions separated by an equals sign. The equals sign provides a balance between the two algebraic expressions. To maintain the balance of an equation, you must make sure that whatever you do to one side you also do to the other side.

$$x + 8 = 11$$
$$x + 8 - 8 = 11 - 8 \qquad \text{Subtracting 8 from both sides.}$$
$$x = 3$$

$$x - 2.5 = 3$$
$$x - 2.5 + 2.5 = 3 + 2.5 \qquad \text{Adding 2.5 to both sides.}$$
$$x = 5.5$$

$$4x = 12$$
$$\frac{4x}{4} = \frac{12}{5} \qquad \text{Dividing both sides by 4.}$$
$$x = 3$$

$$\frac{x}{5} = 3.2$$
$$\frac{x}{5} \times 5 = 3.2 \times 5 \qquad \text{Multiplying both sides by 5.}$$
$$x = 16$$

Worked example

Solve the following equations.

a) $4x + 5 = 17$ **b)** $9x - 1 = 7x + 14$ **c)** $6(2x + 5) = 48$

a)
$$4x + 5 = 17$$
$$4x = 17 - 5 \qquad \text{Subtracting 5 from both sides.}$$
$$4x = 12$$
$$x = \frac{12}{4} \qquad \text{Dividing both sides by 4.}$$
$$x = 3$$

b)
$$9x - 1 = 7x + 14 \qquad \text{This equation has the unknown on both sides.}$$
$$9x = 7x + 14 + 1 \qquad \text{Adding 1 to both sides.}$$
$$9x = 7x + 15$$
$$9x - 7x = 15 \qquad \text{Subtracting 7x from both sides.}$$
$$2x = 15$$
$$x = \frac{15}{2} \qquad \text{Dividing both sides by 2.}$$
$$x = 7\tfrac{1}{2} \text{ or } 7.5$$

c)
$$6(2x + 5) = 48$$
$$6 \times 2x + 6 \times 5 = 48 \qquad \text{Expanding the brackets.}$$
$$12x + 30 = 48$$
$$12x = 48 - 30 \qquad \text{Subtracting 30 from both sides.}$$
$$12x = 18$$
$$x = \frac{18}{12} \qquad \text{Dividing both sides by 12.}$$
$$x = 1\tfrac{1}{2} \text{ or } 1.5$$

You can apply the same method to solve equations in a variety of number problems.

Worked example

Find three consecutive numbers with a sum of 15.

To solve a problem like this, call one of the numbers x. Then, as the numbers are consecutive, the other two numbers are $(x + 1)$ and $(x + 2)$.

If the sum is 15 then $x + (x + 1) + (x + 2) = 15$.

$$x + x + 1 + x + 2 = 15$$

$3x + 3 = 15$	Collecting like terms.
$3x = 15 - 3$	Subtracting 3 from both sides.
$3x = 12$	
$x = \frac{12}{3}$	Dividing both sides by 3.
$x = 4$	

Having found the answer $x = 4$, you need to interpret it. Since the consecutive numbers were x, $(x + 1)$ and $(x + 2)$ then the required numbers are 4, $(4 + 1)$ and $(4 + 2)$ or 4, 5 and 6.

✛ HINT

▸ You should now check your solutions to make sure that they are correct.

Worked example

A group of people shared a pay-out of £45 000 equally, and each of them received £5000. How many people shared the money?

You need to form an equation and solve it. Let the number of people be x.

$\frac{45\,000}{x} = 5000$	
$45\,000 = 5000x$	Multiplying both sides by x.
$\frac{45\,000}{5000} = x$	Dividing both sides by 5000.
$9 = x$	
$x = 9$	Reversing the answer to get x on the left.

Nine people shared the pay-out.

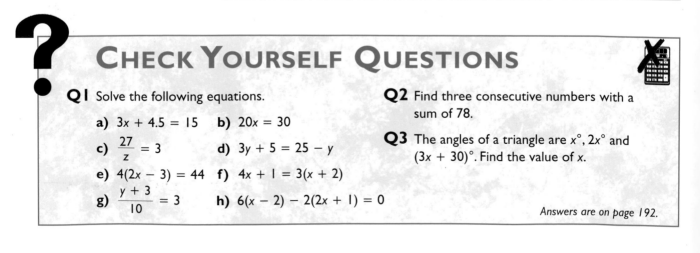

? CHECK YOURSELF QUESTIONS

Q1 Solve the following equations.

a) $3x + 4.5 = 15$ **b)** $20x = 30$

c) $\frac{27}{z} = 3$ **d)** $3y + 5 = 25 - y$

e) $4(2x - 3) = 44$ **f)** $4x + 1 = 3(x + 2)$

g) $\frac{y + 3}{10} = 3$ **h)** $6(x - 2) - 2(2x + 1) = 0$

Q2 Find three consecutive numbers with a sum of 78.

Q3 The angles of a triangle are $x°$, $2x°$ and $(3x + 30)°$. Find the value of x.

Answers are on page 192.

≫ What are the rules?

You can rearrange (or **transpose**) a formula in exactly the same way as you solve an equation. To maintain the balance of the formula you must make sure that whatever you do to one side of the formula you also do to the other side of the formula.

≫ Using the rules

For the formula $S = \frac{D}{T}$, S is called the **subject** of the formula. The formula can be rearranged to make D or T the subject as follows.

$$S = \frac{D}{T}$$

$S \times T = D$ Multiplying both sides of the formula by T.

$D = S \times T$ or $D = ST$ Turning the formula round so that D is the subject of the formula.

Or, from $D = ST$:

$$\frac{D}{S} = T$$ Dividing both sides of the formula by S.

$$T = \frac{D}{S}$$ Turning the formula around so that T is the subject.

Worked example

Make x the subject of the formula $y = 3x + 2$.

$$y = 3x + 2$$

$y - 2 = 3x$ Subtracting 2 from both sides.

$\frac{y - 2}{3} = x$ Dividing both sides of the formula by 3.

$x = \frac{y - 2}{3}$ Turning the formula around so that x is the subject.

? CHECK YOURSELF QUESTIONS

Q1 Rewrite the following with the letter indicated in brackets as the subject.

  a) $C = 2\pi r$ (r) **b)** $v = u + at$ (u)

  c) $v = u + at$ (a) **d)** $A = \pi r^2$ (r)

  e) $V = \pi r^2 h$ (h) **f)** $V = \pi r^2 h$ (r)

  g) $I = \frac{PRT}{100}$ (T)

Q2 Rearrange the formula $a = \frac{m}{b^2}$ to make b the subject.

Answers are on page 193.

Algebraic indices

≫ What are the rules?

From Number, Revision session 5, *Positive, negative and zero indices* you should recall that, in general:

- $a^m \times a^n = a^{m+n}$

- $a^m \div a^n = a^{m-n}$

- $a^{-m} = \dfrac{1}{a^m}$

- $a^0 = 1$ Remember: any number to the power 0 is 1.

≫ Using the rules

Worked example

Simplify these expressions.

a) $a^2 \times a^5$ **b)** $b^7 \div b^7$ **c)** $c^4 \times c \times c^{11}$

a) $a^2 \times a^5 = a^{2+5} = a^7$

b) $b^7 \div b^7 = b^{7-7} = b^0 = 1$

c) $c^4 \times c \times c^{11} = c^{4+1+11} = c^{16}$ As $c = c^1$.

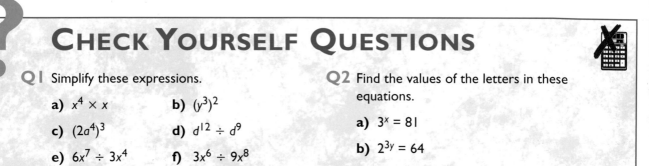

CHECK YOURSELF QUESTIONS

Q1 Simplify these expressions.

   a) $x^4 \times x$ **b)** $(y^3)^2$

   c) $(2a^4)^3$ **d)** $d^{12} \div d^9$

   e) $6x^7 \div 3x^4$ **f)** $3x^6 \div 9x^8$

Q2 Find the values of the letters in these equations.

   a) $3^x = 81$

   b) $2^{3y} = 64$

   c) $5^{2x+1} = 125$

Answers are on page 193.

Interpreting graphs

≫ What do I need to know?

You need to understand how to read graphs and be familiar with interpreting them. They can be presented in many different contexts.

≫ Conversion graphs

Worked example

Draw a conversion graph to show the relationship between miles and kilometres, given that a distance of 5 miles is approximately 8 kilometres.

Use your graph to find:

a) how many kilometres there are in 15 miles

b) how many kilometres there are in 8 miles

c) how many miles there are in $32\frac{1}{2}$ kilometres.

The first requirement is to draw the graph, using the fact that 5 miles is approximately 8 kilometres. This means that a distance of

5 miles is approximately 8 kilometres

10 miles is approximately 16 kilometres

20 miles is approximately 32 kilometres   etc.

As this is a straight-line graph, you should just need three points (two points for the line and one point as a check) to draw the graph. It is also helpful to note that 0 miles is equal to 0 kilometres so the straight line should pass through the origin.

The question asks you to 'use your graph' to find answers to the questions asked and so it is important to show the examiner how you arrived at your solutions and convince them that you did not just calculate the answers.

From the graph:

a) there are approximately 24 kilometres in 15 miles

b) there are approximately 12.8 kilometres in 8 miles
(Remembering that each small square is 1 kilometre on this scale.)

c) there are approximately 20.3 miles in $32\frac{1}{2}$ kilometres.
(Remembering that each small square is 1 mile on this scale.)

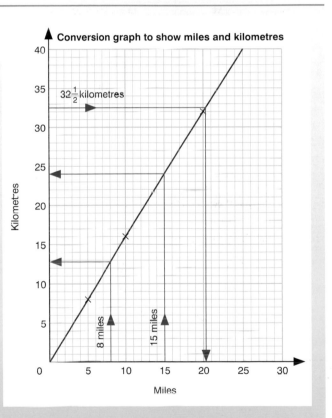

Conversion graph to show miles and kilometres

≫ Travel graphs

Worked example

A salesperson leaves home at 1030 hours and their distance from home is shown on the graph below.

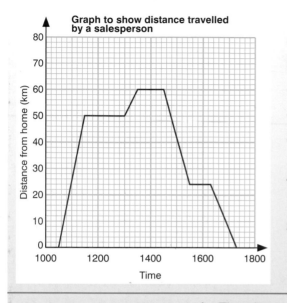

Graph to show distance travelled by a salesperson

Use the graph to answer the following questions.

a) How many kilometres are travelled before the first stop?

b) How long does it take to reach the first stop?

c) How far is the salesperson away from home at 1600 hours?

d) What time does the salesperson arrive back at home?

e) How far does the salesperson travel between 1200 hours and 1600 hours?

f) What is the average speed of the salesperson between 1030 hours and 1130 hours?

g) What is the average speed of the salesperson between 1300 hours and 1330 hours?

a) The salesperson travels 50 kilometres before the first stop.

b) The first stop is after $1\frac{1}{2}$ hours (from 1130 hours to 1300 hours).

c) The salesperson is 24 kilometres away from home at 1600 hours.

d) The salesperson arrives back home at 1718 hours. (Remembering that each small square represents $\frac{1}{5}$ hour or 12 minutes.)

e) The salesperson travels 10 km + 36 km = 46 km.

f) Between 1030 hours and 1130 hours:

distance travelled = 50 km

time taken = 1 hour

so speed = 50 kilometres per hour

(speed = distance ÷ time)

g) Between 1300 hours and 1330 hours:

distance travelled = 10 km

time taken = $\frac{1}{2}$ hour

so speed = 20 kilometres per hour   (speed = distance ÷ time)

Q1 The cooking instructions for a piece of meat are given as:

40 minutes per kilogram plus 25 minutes.

Draw a graph and use it to find:

a) how long a piece of meat weighing 2 kilograms would take to cook

b) how long a piece of meat weighing 1.4 kilograms would take to cook

c) the weight of a piece of meat which takes $1\frac{1}{2}$ hours to cook.

Q2 The graph below shows the journeys of two cyclists travelling between two places that are 25 miles apart.

Use the graph to answer the following questions.

a) How many miles does the first cyclist travel before the first stop?

b) What is the average speed of the first cyclist over this first part of the journey?

c) How long does this first cyclist stop?

d) At what time does the second cyclist overtake the first cyclist?

e) What is the average speed of the second cyclist at this time?

f) What time does the second cyclist arrive at the destination?

g) What is the greatest distance between the two cyclists?

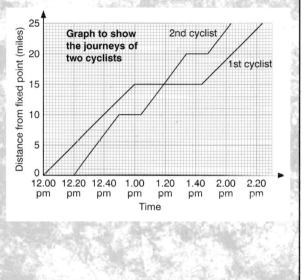

Answers are on page 194.

Linear graphs and coordinates

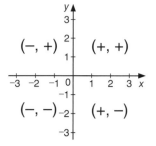

≫ What do the words mean?

Coordinates are used to locate points on a graph.

Negative coordinates can be used by extending the *x*-axis and *y*-axis in the negative directions, dividing the graph into four quadrants.

A **linear graph** is one in which the points can be joined to give a straight line.

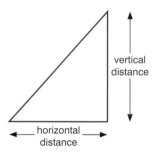

≫ Linear graphs

Linear graphs are straight lines.

The gradient of the line is defined as:

$$\frac{\text{vertical distance}}{\text{horizontal distance}}$$

Gradients can be either positive or negative, depending on their direction of slope.

positive gradient   negative gradient

Worked example

Calculate the gradients of the lines joining the points:
a) $(3, 2)$ and $(7, 10)$ **b)** $(^-8, 5)$ and $(2, ^-1)$.

a) Sketch the two points $(3, 2)$ and $(7, 10)$ on a graph.

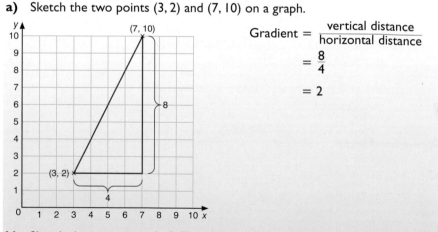

$$\text{Gradient} = \frac{\text{vertical distance}}{\text{horizontal distance}}$$

$$= \frac{8}{4}$$

$$= 2$$

+ HINT

▸ Parallel lines have the same gradient, and lines with the same gradient are parallel.

b) Sketch the two points $(^-8, 5)$ and $(2, ^-1)$ on a graph.

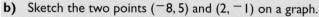

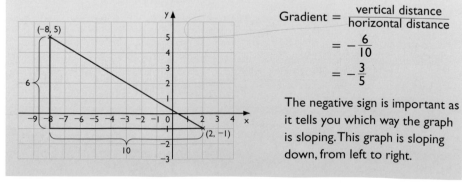

$$\text{Gradient} = \frac{\text{vertical distance}}{\text{horizontal distance}}$$

$$= -\frac{6}{10}$$

$$= -\frac{3}{5}$$

The negative sign is important as it tells you which way the graph is sloping. This graph is sloping down, from left to right.

All linear graphs can be written in the form $y = mx + c$ where m is the **gradient** of the line and c is the cut-off on the y-axis. This is where the line **intersects** the y-axis (also called the y-**intercept**).

Worked example

Sketch the following straight-line graphs.

a) $y = 2x - 1$ **b)** $y + 3x = 3$

You can easily sketch the graphs if you compare their equations with the general form $y = mx + c$ where m is the gradient and c is the cut-off on the y-axis (i.e. the y-intercept).

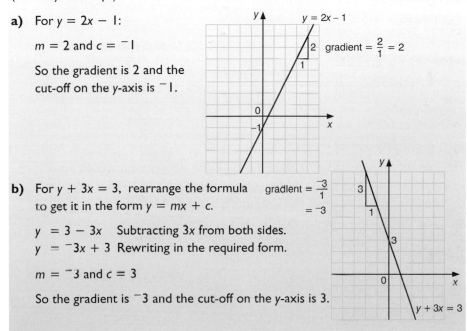

a) For $y = 2x - 1$:

$m = 2$ and $c = {}^-1$

So the gradient is 2 and the cut-off on the y-axis is ${}^-1$.

b) For $y + 3x = 3$, rearrange the formula to get it in the form $y = mx + c$.

$y = 3 - 3x$ Subtracting $3x$ from both sides.
$y = {}^-3x + 3$ Rewriting in the required form.

$m = {}^-3$ and $c = 3$

So the gradient is ${}^-3$ and the cut-off on the y-axis is 3.

? CHECK YOURSELF QUESTIONS

Q1 On the same set of axes, sketch the following lines.

   a) $y = \frac{1}{2}x + 5$ **b)** $y = \frac{1}{2}x + 3$

   c) $y = \frac{1}{2}x - 2$

   What can you say about the lines?

Q2 On the same set of axes, sketch the following graphs.

   a) $y + 3x = 5$ **b)** $x = 2y + 6$

Q3 Write down the equations of the straight-line graphs shown on this grid.

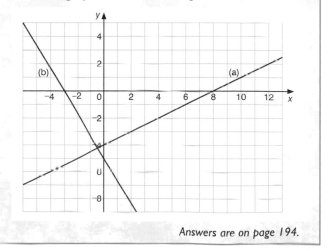

Answers are on page 194.

ALGEBRA

Inequalities and graphs

≫ What do the symbols mean?

The solution of the inequality $x < 5$ can take many values such as 4, π, $\sqrt{2}$, $2\frac{1}{2}$, 0, $-3\frac{1}{4}$, $-100\,000$. You can show such inequalities on a number line.

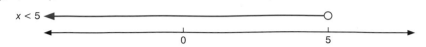

The open circle, ○, at the end of the line shows that the value 5 is not included as $x < 5$.

The solid circle, ●, is used to show that the number is included, as shown below for $x \leqslant 5$.

≫ Solving inequalities

You can solve inequalities in exactly the same way as you would equalities (i.e. equations) except that **when you multiply or divide by a negative number you must reverse the inequality sign.**

Worked example

Solve the following inequalities and show each solution on a number line.

a) $3y + 5 < 17$

b) $^-10c > 5$

c) $5 - 8m \leqslant 13$

d) $4x < 5x + 2$

e) $2 \leqslant \frac{2}{3}(x + 5) \leqslant 6$

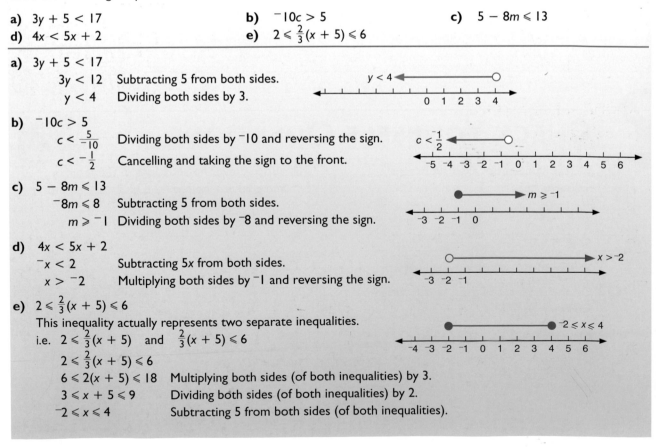

a) $3y + 5 < 17$
$\quad 3y < 12$ Subtracting 5 from both sides.
$\quad y < 4$ Dividing both sides by 3.

b) $^-10c > 5$
$\quad c < \frac{5}{-10}$ Dividing both sides by $^-10$ and reversing the sign.
$\quad c < -\frac{1}{2}$ Cancelling and taking the sign to the front.

c) $5 - 8m \leqslant 13$
$\quad ^-8m \leqslant 8$ Subtracting 5 from both sides.
$\quad m \geqslant ^-1$ Dividing both sides by $^-8$ and reversing the sign.

d) $4x < 5x + 2$
$\quad ^-x < 2$ Subtracting 5x from both sides.
$\quad x > ^-2$ Multiplying both sides by $^-1$ and reversing the sign.

e) $2 \leqslant \frac{2}{3}(x + 5) \leqslant 6$
This inequality actually represents two separate inequalities.
i.e. $2 \leqslant \frac{2}{3}(x + 5)$ and $\frac{2}{3}(x + 5) \leqslant 6$
$\quad 2 \leqslant \frac{2}{3}(x + 5) \leqslant 6$
$\quad 6 \leqslant 2(x + 5) \leqslant 18$ Multiplying both sides (of both inequalities) by 3.
$\quad 3 \leqslant x + 5 \leqslant 9$ Dividing both sides (of both inequalities) by 2.
$\quad ^-2 \leqslant x \leqslant 4$ Subtracting 5 from both sides (of both inequalities).

≫ Graphing inequalities

You can show inequalities on a graph by replacing the inequality sign by an equals (=) sign and drawing the line. This will divide the graph into two regions. You need to decide which of these regions is required.

You should usually shade out the region that is not required, although some examination questions ask you to shade the required region. You must make it clear to the examiner which is your required region, by labelling it as appropriate.

You also need to make clear whether the line is included (i.e. the inequality is ⩽ or ⩾), or excluded (i.e. the inequality is < or >). Use a solid line if the line is included, or a dotted line if it is not included.

Worked example

Draw graphs of these lines.

$x = 2$ $y = 4$ $y = 6 - x$

On your graph, label the region where the points (x, y) satisfy the inequalities:

$x \geqslant 2$ $y < 4$ $y \leqslant 6 - x$.

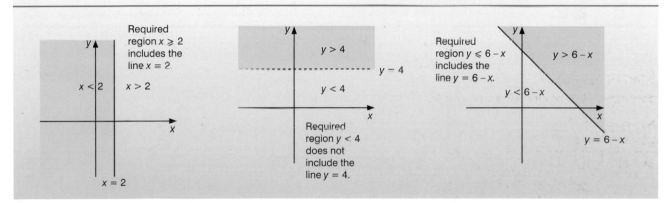

Now combining the graphs:

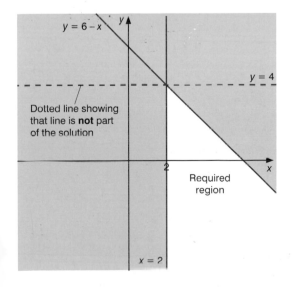

CHECK YOURSELF QUESTIONS

Q1 Solve these inequalities.

  a) $4x + 2 \leqslant 17 - x$

  b) $18 - 6x > 3 - 3x$

Q2 Write down the inequalities illustrated in the unshaded parts of the diagrams below.

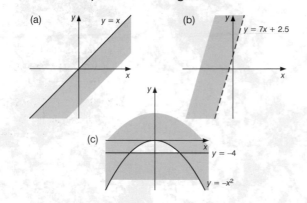

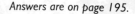

Q3 Draw a set of axes, with each axis labelled from 0 to 10. Label the region represented by these inequalities.

  $x \geqslant 2 \qquad y \geqslant 1 \qquad x + y \leqslant 9$

What is the maximum value of $x + y$ which satisfies all of these conditions?

Answers are on page 195.

Quadratic graphs

≫ What do the graphs look like?

Quadratic graphs all have the same basic shape, as illustrated below.

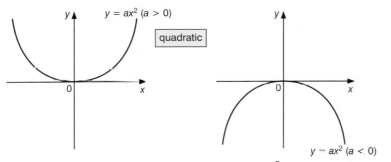

$y = ax^2 \ (a > 0)$

quadratic

$y = ax^2 \ (a < 0)$

Quadratic graphs can be written in the form $y = ax^2 + bx + c$ (where a is non-zero). They all have approximately the same basic shape.

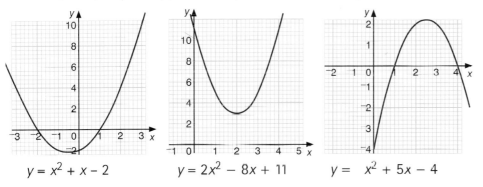

$y = x^2 + x - 2$ $y = 2x^2 - 8x + 11$ $y = x^2 + 5x - 4$

Worked example

Draw the graph of $y = x^2 + 2x - 8$ and use it to solve the equation $x^2 + 2x - 8 = 0$.

Drawing up a table of values:

x	$^-5$	$^-4$	$^-3$	$^-2$	$^-1$	0	1	2	3
$y = x^2 + 2x - 8$	7	0	$^-5$	$^-8$	$^-9$	$^-8$	$^-5$	0	7
Coordinates	$(^-5,7)$	$(^-4,0)$	$(^-3,^-5)$	$(^-2,^-8)$	$(^-1,^-9)$	$(0,^-8)$	$(1,^-5)$	$(2,0)$	$(3,7)$

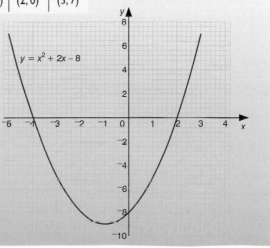

$y = x^2 + 2x - 8$

To solve the equation $x^2 + 2x - 8 = 0$ you need to consider the points that lie on the curve $y = x^2 + 2x - 8$ and on the line $y = 0$. Any points that satisfy both of these equations will also satisfy the equation $x^2 + 2x - 8 = 0$.

From the graph you can see that $x^2 + 2x - 8 = 0$ when the curve crosses the $y = 0$ line, giving $x = {}^-4$ and $x = 2$.

So $x = {}^-4$ and $x = 2$ satisfy the equation $x^2 + 2x - 8 = 0$.

ALGEBRA

Worked example

Draw the graph of $y = 2x^2$ and use it to solve the equation $2x^2 - 5x - 3 = 0$.

Drawing up a table of values:

x	$^-2$	$^-1$	0	1	2	3	4
$y = 2x^2$	8	2	0	2	8	18	32
Coordinates	$(^-2, 8)$	$(^-1, 2)$	$(0, 0)$	$(1, 2)$	$(2, 8)$	$(3, 18)$	$(4, 32)$

To solve the equation $2x^2 - 5x - 3 = 0$ using $y = 2x^2$, rewrite the given equation as follows:

$2x^2 - 5x - 3 = 0$

$2x^2 = 5x + 3$ Adding $5x + 3$ to both sides.

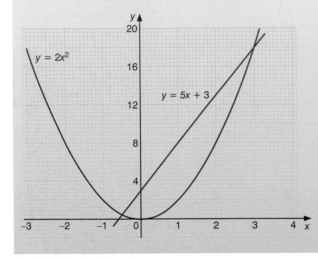

To solve the equation $2x^2 - 5x - 3 = 0$ you need to consider the points that lie on the curve $y = 2x^2$ and on the line $y = 5x + 3$. Any points that satisfy both of these equations will also satisfy the equation $2x^2 - 5x - 3 = 0$.

Draw the graph of $y = 2x^2$ from the table, and then draw the line $y = 5x + 3$.

From the graph you can see that where the two graphs $y = 2x^2$ and $y = 5x + 3$ cross, $x = {}^-\frac{1}{2}$ and $x = 3$, so these values satisfy the equation $2x^2 - 5x - 3 = 0$.

Worked example

Draw the graph of $y = 3x - x^2$ for values of x from $x = {}^-1$ to $x = 4$.

Use your graph to find:

a) the value of x when $3x - x^2$ is as large as possible

b) the values of x for which $3x - x^2$ is greater than 1.5.

To draw the graph of $y = 3x - x^2$ for $x = {}^-1$ to $x = 4$, first draw up a table of values.

x	$^-1$	0	1	2	3	4
$y = 3x - x^2$	$^-4$	0	2	2	0	$^-4$
Coordinates	$(^-1, {}^-4)$	$(0, 0)$	$(1, 2)$	$(2, 2)$	$(3, 0)$	$(4, {}^-4)$

From the graph you can see that:

a) $3x - x^2$ is as large as possible at $x = 1.5$

b) $3x - x^2 = 1.5$ when $x = 0.6$ and when $x = 2.4$.

The values of x for which $3x - x^2$ is greater than 1.5 are $0.6 < x < 2.4$.

Q1 On the same set of axes, draw and label the following graphs.

a) $y = x^2$

b) $y = x^2 + 5$

c) $y = x^2 - 2$

Q2 Draw the graph of $y = x^2 - 6x + 5$.

Use your graph to find:

a) the coordinates of the minimum value of $x^2 - 6x + 5$

b) the values of x when

i) $x^2 - 6x + 5 = 0$
ii) $x^2 - 6x + 5 = 5$.

Q3 The height reached by an object thrown into the air is given by the formula:

$h = 20t - 5t^2$

where h is the height in metres and t is the time in seconds.

Plot the graph of h against t for $0 \leqslant t \leqslant 4$ and use your graph to find the maximum height reached by the object.

Answers are on page 196.

Cubic and reciprocal graphs

≫ What do the graphs look like?

Cubic and reciprocal graphs all have the same basic shapes, as shown below.

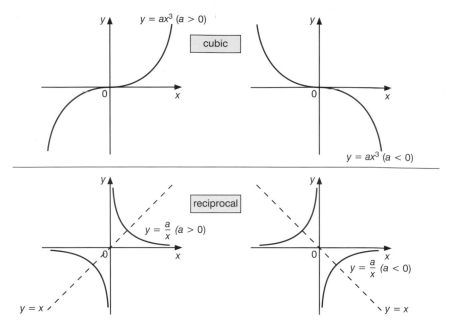

HINT

▸ When drawing cubic and reciprocal graphs it is important that you join the points with a smooth curve rather than a series of straight lines.

≫ Cubic graphs

Cubic graphs can be written in the form $y = ax^3 + bx^2 + cx + d$ (where a is non-zero). They all have approximately the same basic shape as in the examples that follow.

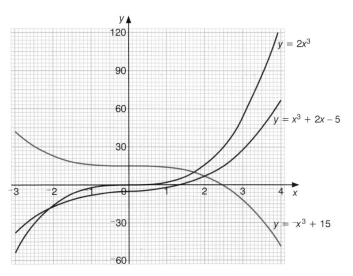

Worked example

Draw the graph of $y = (x - 1)(x - 2)(x - 4)$ for $0 \leqslant x \leqslant 5$ and use it to solve the equation $(x - 1)(x - 2)(x - 4) = 0$.

Drawing up a table of values:

x	0	0.5	1	1.5	2	2.5	3	3.5	4	4.5	5
$y = (x-1)(x-2)(x-4)$	⁻8	⁻2.625	0	0.625	0	⁻1.125	⁻2	⁻1.875	0	4.375	12
Coordinates	(0, −8)	(0.5, ⁻2.625)	(1, 0)	(1.5, 0.625)	(2, 0)	(2.5, ⁻1.125)	(3, ⁻2)	(3.5, ⁻1.875)	(4, 0)	(4.5, 4.375)	(5, 12)

From the graph you can see that $(x - 1)(x - 2)(x - 4) = 0$ when the curve crosses the line $y = 0$, giving $x - 1, x - 2$ and $x = 4$ so that $x = 1, x = 2$ and $x = 4$ satisfy the equation $(x - 1)(x - 2)(x - 4) = 0$.

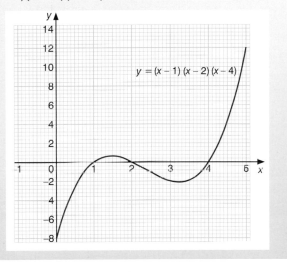

+ HINT

▸ You may need to use non-integer values to find out sufficient detail about the behaviour of the curve and to be able to draw its graph.

≫ Reciprocal graphs

Reciprocal graphs can be written in the form $y = \dfrac{a}{x}$ or $xy = a$ (where a is a constant). They all have approximately the same basic shape as in the examples below.

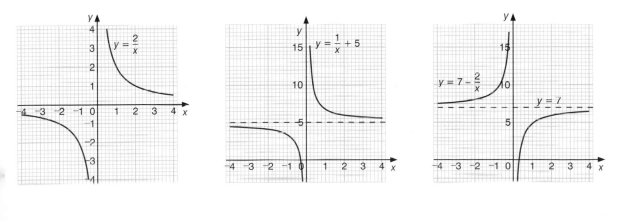

Worked example

On the same set of axes, draw the graph of $xy = 12$ and $y = x$. Use your graphs to solve the equation $x^2 = 12$.

The graph of $xy = 12$ is the same as $y = \dfrac{12}{x}$ (or $x = \dfrac{12}{y}$).

Drawing up a table of values:

x	$^-4$	$^-3$	$^-2$	$^-1$	0	1	2	3	4
$y = x$	$^-4$	$^-3$	$^-2$	$^-1$	0	1	2	3	4
$y = \dfrac{12}{x}$		$^-3$	$^-4$	$^-6$	$^-12$		12	6	4	3

The reciprocal curve $y = \dfrac{12}{x}$ is not defined at $x = 0$.

You will need to use non-integer values, between 0 and 1, and 0 and $^-1$, to find out sufficient detail about the behaviour of the curve and to be able to draw its graph.

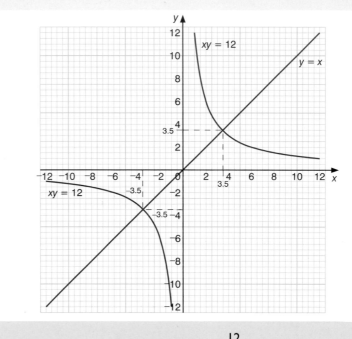

From the graph you can see that the curve $y = \dfrac{12}{x}$ crosses the line $y = x$ when $x \approx {}^-3.5$ and $x \approx 3.5$ correct to 1 d.p.

So $x = {}^-3.5$ and $x = 3.5$ satisfy the equation $\dfrac{12}{x} = x$.

i.e. $12 = x \times x$ Multiplying both sides by x.

or $x^2 = 12$ As required.

HINTS

▶ Since $xy = 12$ and $y = x$, at the point of intersection $x \times x = 12$ which gives $x^2 = 12$, $x = \sqrt{12} \approx 3.5$.

▶ The sign $\approx$ means 'is approximately equal to'.

? CHECK YOURSELF QUESTIONS

Q1 Draw the graph $y = 2x^3 - 3x^2 - 11x + 6$ for $^-2 \leqslant x \leqslant 4$.

Q2 By drawing the graphs of the equation $y = \dfrac{12}{x}$ and $y = x^2 - 1$ on the same axes, solve the equation $\dfrac{1}{x} = x^2 - 1$.

Answers are on page 196.

≫ What are simultaneous equations?

A pair of simultaneous equations is a pair of equations in **two unknowns.**
Both equations are correct at the same time, or simultaneously.
Simultaneous equations are usually solved by **graphical** or **algebraic**
methods. The examination question should make it clear which method you
should use. Otherwise any method will be acceptable.

≫ Graphical solution

To solve the equations graphically, plot the two lines. The coordinates of the
point of intersection give the solution of the simultaneous equations.

Worked example

Solve these simultaneous equations.

$y = x + 4$
$y = 2x + 3$

Plot the simultaneous equations $y = x + 4$
and $y = 2x + 3$ as a pair of straight lines,
like this.

The coordinates of the point of
intersection, $(1, 5)$ give the solution of the
simultaneous equations. At the point of
intersection $(1, 5)$, $x = 1$ and $y = 5$.
So the solution is $x = 1$ and $y = 5$.

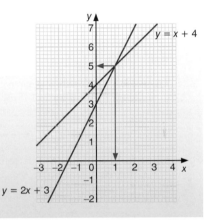

> **+ HINT**
>
> ▸ You should now check this
> solution by substituting the
> values in the original
> equations.

≫ Algebraic solution

There are two algebraic methods for solving simultaneous equations.

SUBSTITUTION METHOD

Rewrite one equation to make one of the unknowns the subject. Then you
can **substitute** this expression into the second equation and solve it.

Worked example

Solve these simultaneous equations.

$x + 2y = 5$
$3x - 2y = 7$

Using $x + 2y = 5$ you can write:

$x = 5 - 2y$ Making x the subject of the equation.

Substitute this value of x into the second equation.

$3x - 2y = 7$	
$3(5 - 2y) - 2y = 7$	Substituting $x = 5 - 2y$.
$15 - 6y - 2y = 7$	Expanding the brackets.
$8 = 8y$	Collecting like terms on each side.
$y = 1$	

Worked example (continued)

Now use $x = 5 - 2y$ with $y = 1$ to find x.

$x = 5 - 2y$

$x = 5 - 2 \times 1$ As $y = 1$.

$x = 3$

ELIMINATION METHOD

Add or subtract the equations, or multiples of them, to **eliminate** one of the unknowns. Then solve the resulting equation.

Worked example

Solve these simultaneous equations.

$x + 2y = 5$
$3x - 2y = 7$

$x + 2y = 5$
$3x - 2y = 7$

If you add the left-hand sides of the equations then y will be eliminated.

The sum of the two left-hand sides must equal the sum of the two right-hand sides.

$(x + 2y) + (3x - 2y) = 5 + 7$ or $x + 2y = 5$
$\quad x + 2y + 3x - 2y = 12 \qquad\qquad + 3x - 2y = 7$
$\qquad\qquad\qquad\quad 4x = 12 \qquad\qquad\quad 4x \quad\;\; = 12$
$\qquad\qquad\qquad\quad\; x = 3 \qquad\qquad\qquad\;\; x = 3$

Substituting for y in the first equation:

$x + 2y = 5$
$3 + 2y = 5$ As $x = 3$.
$\quad\;\; 2y = 2$ Subtracting 3 from both sides.
$\qquad y = 1$ Dividing both sides by 2.

? CHECK YOURSELF QUESTIONS

Q1 Solve the following simultaneous equations by:

    i) the graphical method
    ii) the substitution method
    iii) the elimination method.

a) $x + 3y = 10$
     $2x - 3y = 2$

b) $x = 5y - 3$
     $3x - 8y = 12$

c) $x - 1 = y$
     $3x + 4y = 6$

Q2 Find two numbers with a sum of 36 and a difference of 4.

Q3 The cost of two ties and a shirt is £32.50 while the cost of one tie and two shirts is £41.00. What is the cost of a tie and a shirt?

Answers are on page 197.

Quadratic equations

≫ What do the symbols mean?

Quadratic equations are equations of the form $ax^2 + bx + c = 0$ where $a \neq 0$. You can solve quadratic equations in a number of ways but at this level you would normally use graphical or algebraic methods. The examination question should make it clear which method you should use. If it does not, any method will be acceptable.

See Algebra, Revision session 11, *Quadratic graphs* for further information on how to solve quadratic equations by graphical methods.

≫ Solution by factors

If the product of two numbers is zero then one or both of the numbers must be zero.

If $ab = 0$ then either $a = 0$ or $b = 0$ or both $a = 0$ and $b = 0$.
You can apply this fact to the solution of quadratic equations.

Worked example
Solve the quadratic equation $(x - 3)(x + 1) = 0$.

Since the product of the two factors is zero then one or both of the factors must be zero so:

either $(x - 3) = 0$ which implies that $x = 3$
or $(x + 1) = 0$ which implies that $x = {}^-1$.

The solutions of the equation $(x - 3)(x + 1) = 0$ are $x = 3$ and $x = {}^-1$.

Worked example
Solve the quadratic equation $x^2 + 4x - 21 = 0$.

Factorising the left-hand side of the equation:

$x^2 + 4x - 21 = (x \quad)(x \quad)$ You need to look for numbers which, when multiplied
$\qquad\qquad\quad = (x - 3)(x + 7)$ together, give $^-21$ and which, when added, give 4.
$\qquad\qquad\qquad\qquad$ Try $^+1$ and $^-21$ $^-1$ and $^+21$
$\qquad\qquad\qquad\qquad\qquad$ $^+3$ and $^-7$ $^-3$ and $^+7$

The only pair to satisfy the equation is $^-3$ and $^+7$.

The quadratic equation can be written $(x - 3)(x + 7) = 0$.

Since the product of the two factors is zero then one or both of the factors must be zero so:

either $(x - 3) = 0$ which implies that $x = 3$
or $(x + 7) = 0$ which implies that $x = {}^-7$.

The solutions of the equation $x^2 + 4x - 21 = 0$ are $x = 3$ and $x = {}^-7$.

Q1 Solve the following quadratic equations.

 a) $(x - 5)(x - 7) = 0$

 b) $(x - 6)(2x + 1) = 0$

 c) $x^2 + 4x - 5 = 0$

Q2 Write down quadratic equations with solutions as follows.

 a) $x = 2$ and $x = 5$
 b) $x = {}^-3$ and $x = \frac{1}{5}$

Q3 Solve $x^2 - 5x + 2 = 8$.

Q4 The length of a rectangle is 4 centimetres more than its width. The area of the rectangle is 96 square centimetres. What are the length and the width of the rectangle?

Answers are on page 198.

≫ What is trial and improvement?

You can use trial and improvement to provide successively better **approximations** to the solution of a problem. Your first approximation is repeatedly refined to provide an **improved solution** until you reach the desired accuracy.

You can always find an initial approximation by drawing a graph of the function, although this initial value will usually be given in the question. Alternatively, you could try a few calculations, with whole numbers, in your head.

Worked example

The length of a rectangle is 2 cm greater than the width.
The area of the rectangle is 30 cm^2.
Use trial and improvement to obtain the length and width, to the nearest millimetre.

A useful way to solve this problem is to draw up a table.

Width	Length (width + 2)	Area	Comments	
4	6	24	too small	
5	7	35	too large	Width must lie between 4 and 5.
4.5	6.5	29.25	too small	Width must lie between 4.5 and 5.
4.7	6.7	31.49	too large	Width must lie between 4.5 and 4.7.
4.6	6.6	30.36	too large	Width must lie between 4.5 and 4.6.
4.55	6.55	29.8025	too small	Width must lie between 4.55 and 4.6.

Since 4.55 and 4.6 are both equal to 4.6 (correct to the nearest one-tenth of a centimetre, or millimetre) then you can stop and say that the solution is 4.6 (correct to the nearest millimetre).

Alternatively, notice that:

using 4.5 gives 29.25, which is ⁻0.75 from 30

using 4.6 gives 30.36, which is 0.36 from 30.

So 4.6 is closer.

Worked example

A solution of the equation $x^3 - 3x = 25$ lies between 3 and 4. Use a trial and improvement method to find this solution, giving your answer correct to 1 d.p.

When $x = 3$ $x^3 - 3x = 3^3 - 3 \times 3 = 18$

When $x = 4$ $x^3 - 3x = 4^3 - 3 \times 4 = 52$

So the solution lies between 3 and 4 (and seems to be closer to $x = 3$).

Try $x = 3.5$ $x^3 - 3x = 3.5^3 - 3 \times 3.5 = 32.375$

So the solution lies between 3 and 3.5 (and seems to be closer to $x = 3.5$).

Try $x = 3.3$ $x^3 - 3x = 3.3^3 - 3 \times 3.3 - 26.037$

So the solution lies between 3 and 3.3 (and seems to be closer to $x = 3.3$).

Try $x = 3.2$ $x^3 - 3x = 3.2^3 - 3 \times 3.2 = 23.168$

So the solution lies between 3.2 and 3.3.

Try $x = 3.25$ $x^3 - 3x = 3.25^3 - 3 \times 3.25 = 24.578125$

So the solution lies between 3.25 and 3.3.

Since 3.25 and 3.3 are both equal to 3.3 (correct to 1 decimal place) then you can stop and say that the solution is $x = 3.3$ (correct to 1 decimal place).

Alternatively:

using 3.2 gives 23.168 which is ¯1.832 from 25

using 3.3 gives 26.037 which is 1.037 from 25.

So $x = 3.3$ is the better solution.

CHECK YOURSELF QUESTIONS

Q1 A solution of the equation $x^3 + x = 100$ lies in the range $4 \leqslant x \leqslant 5$. Use trial and improvement to find the solution, correct to 1 decimal place.

Q2 Using the method of trial and improvement, solve the equation $x^3 - x = 5$, correct to 1 decimal place.

Answers are on page 199.

Further algebra

≫ Simplifying expressions

At the Higher level you are required to simplify complicated expressions, solve more difficult equations and rearrange formulae where the subject occurs in more than one term.

Worked example

Simplify fully the expression $\dfrac{x^2 - 16}{2x^2 + 7x - 4}$.

Factorising the numerator: $\quad x^2 - 16 = (x - 4)(x + 4)$

> Remember the difference of two squares, where $x^2 - 16 = x^2 - 4^2$.

Factorising the denominator: $2x^2 + 7x - 4 = (2x - 1)(x + 4)$

Combining these: $\qquad \dfrac{x^2 - 16}{2x^2 + 7x - 4} = \dfrac{(x - 4)(x + 4)}{(2x - 1)(x + 4)}$ Cancelling the $(x + 4)$ factors.

$$= \dfrac{(x - 4)}{(2x - 1)}$$

Worked example

Make x the subject of the formula $p = \dfrac{xy}{x - y}$.

$p = \dfrac{xy}{x - y}$

$p(x - y) = xy$ Multiplying both sides by $(x - y)$.

$px - py = xy$ Expanding the brackets.

$px = xy + py$ Adding py to both sides, to try to isolate terms in x.

$px - xy = py$ Collecting terms in x on one side only.

$x(p - y) = py$ Factorising the left-hand side with x outside the bracket.

$x = \dfrac{py}{p - y}$ Dividing both sides by $(p - y)$ to make x the subject.

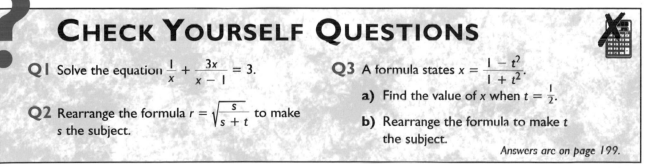

? CHECK YOURSELF QUESTIONS

Q1 Solve the equation $\dfrac{1}{x} + \dfrac{3x}{x - 1} = 3$.

Q2 Rearrange the formula $r = \sqrt{\dfrac{s}{s + t}}$ to make s the subject.

Q3 A formula states $x = \dfrac{1 - t^2}{1 + t^2}$.

   a) Find the value of x when $t = \dfrac{1}{2}$.

   b) Rearrange the formula to make t the subject.

Answers are on page 199.

Further simultaneous equations

≫ What do I need to know?

At the Higher level you are required to solve simultaneous equations where one equation is linear and the other is quadratic, or where one equation is linear and the other is of the form $x^2 + y^2 = r^2$ (where $x^2 + y^2 = r^2$ defines a circle centred on the origin, with a radius of r).

Simultaneous equations can be solved by **graphical** or **algebraic** methods. The examination question should make it clear which method you should use. Otherwise, any method will be acceptable.

≫ Graphical solution

You can solve simultaneous equations graphically by plotting the two equations. The coordinates of the points of intersection give the solutions of the simultaneous equations.

Worked example

Solve these simultaneous equations.

$y = x + 4$
$y = x^2 + 2$

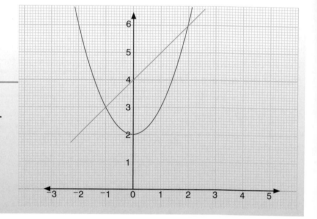

You can plot the simultaneous equations $y = x + 4$ and $y = x^2 + 2$ as a straight line and a curve, as shown on the right.

The coordinates of the points of intersection, $(^-1, 3)$ and $(2, 6)$, give the solutions of the simultaneous equations.

At the points of intersection, $x = {}^-1, y = 3$ and $x = 2, y = 6$.

Worked example

Solve these simultaneous equations.

$y = x + 1$
$x^2 + y^2 = 25$

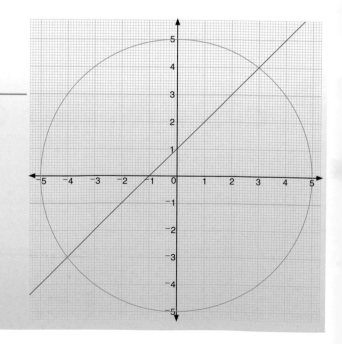

You can plot the simultaneous equations $y = x + 1$ and $x^2 + y^2 = 25$ as a straight line and a circle, like this.

The equation $x^2 + y^2 = 25$ defines a circle, centred on the origin, with radius 5.

The coordinates of the points of intersection, $(3, 4)$, and $(^-4, ^-3)$, give the solutions of the simultaneous equations.

At the points of intersection, $x = 3, y = 4$ and $x = {}^-4, y = {}^-3$.

≫ Algebraic solution

You would generally use the substitution method when solving simultaneous equations of the type shown in the following worked examples.

Worked example

Solve these simultaneous equations.

$y = x + 4$
$y = x^2 + 2$

Use the second equation.

$$y = x^2 + 2$$
$$(x + 4) = x^2 + 2 \qquad \text{Substituting } y = x + 4.$$
$$0 = x^2 + 2 - x - 4 \quad \text{Collecting terms on one side.}$$
$$x^2 - x - 2 = 0 \qquad \text{Simplifying to obtain a quadratic equation.}$$
$$(x + 1)(x - 2) = 0 \qquad \text{Factorising the quadratic.}$$

Either $x = {}^-1$ or $x = 2$.
If $x = {}^-1$ then $y = 3$.   As $y = x + 4$.
If $x = 2$ then $y = 6$.   Again, as $y = x + 4$.

The solutions are $x = {}^-1, y = 3$ and $x = 2, y = 6$.

Worked example

Solve these simultaneous equations.

$y = x + 1$
$x^2 + y^2 = 25$

Use the second equation.

$$x^2 + y^2 = 25$$
$$x^2 + (x + 1)^2 = 25 \qquad \text{Substituting } y = x + 1.$$
$$x^2 + x^2 + 2x + 1 = 25 \qquad \text{Expanding the brackets.}$$
$$2x^2 + 2x - 24 = 0 \qquad \text{Collecting the terms on one side.}$$
$$x^2 + x - 12 = 0 \qquad \text{Dividing both sides by 2.}$$
$$(x + 4)(x - 3) = 0 \qquad \text{Factorising the quadratic.}$$

Either $x = {}^-4$ or $x = 3$.
If $x = {}^-4$ then $y = {}^-3$.   As $y = x + 1$.
If $x = 3$ then $y = 4$.   Again, as $y = x + 1$.

The solutions of the simultaneous equations are $x = {}^-4, y = {}^-3$ and $x = 3, y = 4$.

? CHECK YOURSELF QUESTIONS

Q1 By drawing suitable graphs, solve the following pairs of simultaneous equations.

    a) $y = 11x - 2$ **b)** $x + y = 7$
        $y = 5x^2$ $x^2 + y^2 = 25$

Q2 Use an algebraic method to solve the simultaneous equations given in question 1.

Answers are on page 200.

Further quadratic equations

≫ What do I need to know?

At the Higher level you will be expected to solve quadratic equations by a variety of methods including **using the formula** and **iteration techniques**. The examination question should make it clear which method you should use or else any method will be acceptable

See Algebra, Revision session 11, *Quadratic graphs* for further information on how to solve quadratic equations by graphical methods and see Algebra, Revision session 14, *Quadratic equations* for further information on factorising.

≫ Solving quadratic equations

There are two methods for solving quadratic equations of the form $ax^2 + bx + c \, (a \neq 0)$.

COMPLETING THE SQUARE

This method of solving quadratic equations involves forming a 'perfect square' of the form $(x + \frac{b}{2a})^2 + \frac{c}{2} - (\frac{b}{2a})^2 = 0$, then taking square roots, as illustrated in the following examples.

Worked example

Solve the equation $x^2 + 4x + 3 = 0$.

Comparing the given quadratic with the general form $ax^2 + bx + c = 0$:

$a = 1, b = 4, c = 3$.

Since $a = 1$, the 'square' term is $(x + \frac{b}{2})^2$.

The 'square' is $(x + \frac{4}{2})^2$ or $(x + 2)^2$.

Now	$(x + 2)^2 = x^2 + 4x + 4$	
or	$x^2 + 4x + 4 = (x + 2)^2$	Turning the equation around.
and	$x^2 + 4x = (x + 2)^2 - 4$	Subtracting 4 from both sides.
But	$(x^2 + 4x) + 3 = 0$	Using the original quadratic.
So	$[(x + 2)^2 - 4] + 3 = 0$	Replacing $x^2 + 4x$ by $(x + 2)^2 - 4$.
	$(x + 2)^2 - 1 = 0$	Simplifying.
	$(x + 2)^2 = 1$	Isolating the 'square' term on the left.
	$(x + 2) = \sqrt{1}$	Taking square roots on both sides.
	$(x + 2) = \pm 1$	The square root has two solutions.
So	$(x + 2) = {}^+1$ and $x = {}^-1$	
or	$(x + 2) = {}^-1$ and $x = {}^-3$	

In this case, the quadratic could have been factorised quite simply. However, this is not the case in the next example.

Worked example

Solve the equation $x^2 + 3x - 2 = 0$, leaving your answer in surd form.

Comparing the given quadratic with the general form $ax^2 + bx + c = 0$:

$a = 1, b = 3, c = {}^-2$.

Since $a = 1$, the 'square' term is $(x + \frac{b}{2})^2$.

The 'square' is $(x + \frac{3}{2})^2$.

Now	$(x + \frac{3}{2})^2 = x^2 + 3x + \frac{9}{4}$	
or	$x^2 + 3x + \frac{9}{4} = (x + \frac{3}{2})^2$	Turning the equation around.
and	$x^2 + 3x = (x + \frac{3}{2})^2 - \frac{9}{4}$	Subtracting $\frac{9}{4}$ from both sides.
But	$(x^2 + 3x) - 2 = 0$	Using the original quadratic.
So	$[(x + \frac{3}{2})^2 - \frac{9}{4}] - 2 = 0$	Replacing $x^2 + 3x$ by $(x + \frac{3}{2})^2 - \frac{9}{4}$.
	$(x + \frac{3}{2})^2 - \frac{17}{4} = 0$	Simplifying.
	$(x + \frac{3}{2})^2 = \frac{17}{4}$	Isolating the square term on the LHS.
	$(x + \frac{3}{2}) = \sqrt{\frac{17}{4}}$	Taking square roots on both sides.
	$(x + \frac{3}{2}) = \pm\frac{\sqrt{17}}{2}$	The square root has two solutions.
So	$(x + \frac{3}{2}) = +\frac{\sqrt{17}}{2}$ and $x = -\frac{3}{2} + \frac{\sqrt{17}}{2} = \frac{-3 + \sqrt{17}}{2}$ or $\frac{\sqrt{17} - 3}{2}$	
or	$(x + \frac{3}{2}) = -\frac{\sqrt{17}}{2}$ and $x = -\frac{3}{2} - \frac{\sqrt{17}}{2} = \frac{-3 - \sqrt{17}}{2}$ or $-\left(\frac{3 + \sqrt{17}}{2}\right)$	

USING THE FORMULA

As an alternative to completing the square, you can use the formula:

$$x = \frac{-b \pm \sqrt{b^2 - 4ac}}{2a}$$

to solve any quadratic of the form $ax^2 + bx + c = 0$.

Worked example

Solve the equation $x^2 + 3x - 2 = 0$, giving your answer to an appropriate degree of accuracy.

Comparing the given quadratic with the general form $ax^2 + bx + c = 0$:

$a = 1, b = 3, c = {}^-2$.

Substituting these values in the formula:

$$x = \frac{{}^-b \pm \sqrt{b^2 - 4ac}}{2a}$$

$$x = \frac{{}^-3 \pm \sqrt{3^2 - 4 \times 1 \times {}^-2}}{2 \times 1}$$

$$x = \frac{{}^-3 \pm \sqrt{9 - {}^-8}}{2}$$

$$x = \frac{{}^-3 \pm \sqrt{17}}{2}$$

$$x = \frac{{}^-3 \pm 4.123\,105\,626}{2}$$

+ **HINT**

▶ This formula for solving quadratic equations will be given in the examination so you do not need to memorise it but you do need to know how to use it correctly.

+ **HINT**

▶ The answer $\frac{3 \pm \sqrt{17}}{2}$ is the same as the answer obtained in the example above.

ALGEBRA

$$x = \frac{-3 + 4.123\,105\,626}{2} \quad \text{or} \quad x = \frac{-3 - 4.123\,105\,626}{2}$$

$$x = \frac{1.123\,105\,626}{2} \quad \text{or} \quad x = \frac{-7.123\,105\,626}{2}$$

$$x = 0.561\,552\,813 \quad \text{or} \quad x = -3.561\,552\,813$$

$$x = 0.562 \text{ (3 s.f.)} \quad \text{or} \quad x = -3.56 \text{ (3 s.f.)}$$

SOLUTION BY ITERATION

Iteration involves using an initial value (usually denoted by x_1) to find successive solutions (x_2, x_3, x_4, ...). Each solution is based on the previous solution to improve the accuracy, so that x_{n+1} is calculated from x_n, using the given iterative formula.

Iterative formulae can be used as an alternative method for solving quadratic equations. The iterative formula can be found by rearranging the equation.

Worked example

Solve the quadratic equation $x^2 - 5x + 1 = 0$.

$x^2 - 5x + 1 = 0$

$x^2 + 1 = 5x$ Adding $5x$ to both sides.

$5x = x^2 + 1$ Turning the equation around.

$x = \frac{1}{5}(x^2 + 1)$

Now, writing this as an iterative formula:

$x_{n+1} = \frac{1}{5}(x_n^2 + 1)$

Any equation can give rise to a number of different iterative formulae, although they are not always of any use in solving the quadratic. In the above example $x_{n+1} = \frac{1}{5}(x_n^2 + 1)$ provides a solution to the quadratic whereas $x_{n+1} = 5 - \dfrac{1}{x_n}$ will not, as successive iterations will diverge.

? CHECK YOURSELF QUESTIONS

Q1 Solve the equation $2x^2 + 5x + 2 = 0$ by completing the square.

Q2 Solve the following quadratic equations by using the formula $x = \dfrac{-b \pm \sqrt{b^2 - 4ac}}{2a}$.

a) $x^2 + 2x - 1 = 0$ **b)** $2x^2 = 10x - 4$

Q3 The length of a room is 4 metres longer than its width. Find the dimensions of the room if the area is 32 square metres.

Q4 Use the iteration $x_{n+1} = 10 + \dfrac{1}{x_n}$ with $x_1 = 5$ to find a root of the equation $x^2 - 10x - 1 = 0$ correct to four decimal places.

Q5 A sequence is given by $x_{n+1} = \dfrac{6}{x_n + 5}$. The first term, x_1, of the sequence is 3.

a) Find the next three terms.

b) What do you think is the value of x_n as n becomes very large? Write down this value.

c) Show that the quadratic equation which the sequence above is intended to solve is $x^2 + 5x - 6 = 0$.

d) Solve this quadratic equation.

Answers are on page 200.

Gradients and tangents

HIGHER

≫ What do I need to know?

At the Higher level you are required to know that the **gradients of perpendicular lines are negative reciprocals** of each other. You must be able to **use a tangent to calculate the gradient of a curve**, and **interpret the gradient of a graph and the area under the graph.**

≫ Gradients of perpendicular lines

The equation of any linear graph can be written in the form $y = mx + c$, where m is the gradient of the line and c is where the line intersects the y-axis (also called the y-intercept).

The gradients of perpendicular lines are **negative reciprocals** of each other. This means that for two perpendicular lines, $y = m_1x + c_1$ and $y = m_2x + c_2$, then $m_1m_2 = {}^-1$.

Conversely, lines with gradients that are negative reciprocals are perpendicular.

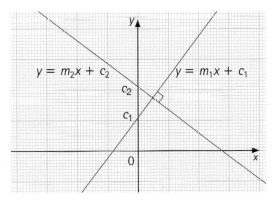

Worked example

Find the equation of the line which is perpendicular to $y = 3x - 2$ and passes through the point $(6, 2)$.

It might be helpful to draw a diagram, to solve the question.

The gradient of the line $y = 3x - 2$ is 3 so the gradient of the line perpendicular to it is the negative reciprocal of 3. The gradient of the perpendicular line is $-\frac{1}{3}$ (since $3 \times {}^-\frac{1}{3} = {}^-1$).

Therefore, this line has the equation $y = {}^-\frac{1}{3}x + c$.

Since this line passes through $(6, 2)$, you can use this to work out the value of c.

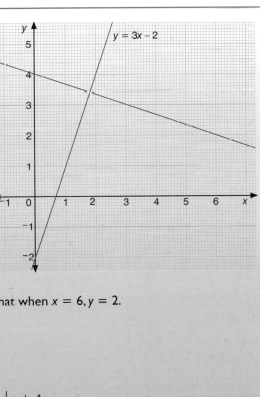

$y = {}^-\frac{1}{3}x + c$

$2 = {}^-\frac{1}{3} \times 6 + c$ Using the fact that when $x = 6, y = 2$.

$2 = {}^-2 + c$

$4 = c$

$c = 4$

So the equation of the line is $y = \frac{1}{3}x + 4$.

> **+ HINT**
>
> ▸ $m_1m_2 = {}^-1$
>
> means $m_1 = -\dfrac{1}{m_2}$
>
> and $m_2 = -\dfrac{1}{m_1}$
>
> which is why they are called negative reciprocals.

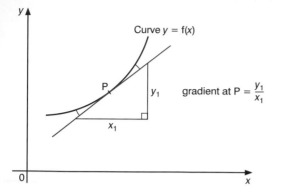

≫ Gradient of a curve

Unlike the gradient of a straight line, **the gradient of a curve changes,** so it will be different depending upon where you want to calculate it. You can find the gradient at a point on a curve by drawing a tangent at that point and working out the gradient as before.

To draw a tangent to a curve you place your ruler on the curve at the required point so that the angles produced at either side are approximately equal.

Worked example

Find the gradient at the following points for the curve $y = 10x - x^2$.

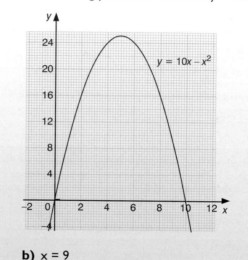

a) $x = 3$ **b)** $x = 9$

c) Write down the coordinates of the point where the gradient $= 0$.

a) To find the gradient at $x = 3$ you need to draw a tangent at the point and work out the gradient.

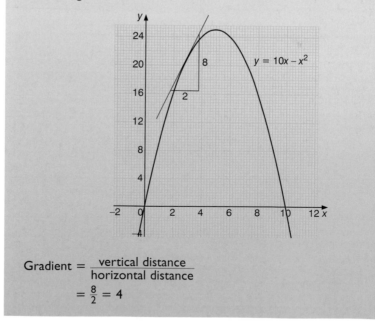

$$\text{Gradient} = \frac{\text{vertical distance}}{\text{horizontal distance}}$$

$$= \frac{8}{2} = 4$$

b) To find the gradient at $x = 9$ you need to draw a tangent at the point and work out the gradient.

$$\text{Gradient} = \frac{\text{vertical distance}}{\text{horizontal distance}}$$

$$= \frac{-8}{2} = {}^-8 \text{ As the gradient is negative.}$$

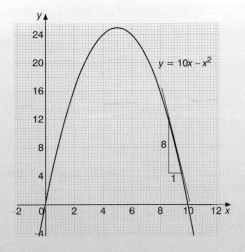

c) From the graph you can see that the gradient is 0 when $x = 5$. You can find the y-value at this point by substituting $x = 5$ in the equation $y = 10x - x^2$.

$$y = 10x - x^2 = 10 \times 5 - 5^2 = 25$$

So the coordinates when the gradient is 0 are $(5, 25)$.

+ HINT

▸ The point at which the gradient $= 0$ is sometimes called the turning point and represents the position of a maximum or minimum value for the function. In this case the maximum value of $y = 10x - x^2$ is 25.

? CHECK YOURSELF QUESTIONS

Q1 Find the equation of the line which is perpendicular to $y = {}^-5x + 7$ and crosses the x-axis at $x = 2$.

Q2 Find the gradient of the curve $y = x^2$ at the following points.

a) $x = 2$ **b)** $x = {}^-2$

c) $x = 5$ **d)** $x = 0$

Q3 Find the gradient of the curve $y = x^3 + 2$ at the following points.

a) $x - 2$ **b)** $x = 4$

c) $x = {}^-1$ **d)** $x = {}^-3$

Answers are on page 202.

Further functions and graphs

≫ What do I need to know?

At the Higher level, you will need to use **function notation**. You will also need to be familiar with **transformations of graphs**.

≫ Function notation

Function notation is a useful tool for describing the relationship between two variables. In function notation the relationship $y = \ldots$ is written as $f(x) = \ldots$ so that the relationship that can be expressed as $y = ax^2 + bx + c$ is written $f(x) = ax^2 + bx + c$ where $y = f(x)$.

It is important to appreciate that $f(x)$ does not mean f multiplied by x but it is a short way of writing 'the function of x'.

If $f(x) = x^2 - 4x + 3$ then f(5) is the value of the function when $x = 5$ and $f(^-2)$ is the value of the function when $x = ^-2$. In this case:

$f(5) = 5^2 - 4 \times 5 + 3 = 8$ and

$f(^-2) = (^-2)^2 - 4 \times {}^-2 + 3 = 15$

etc.

≫ Transformations

There are four different graph transformations with which you need to be familiar. When using these transformations it is helpful to try out a few points to check that you have the correct idea. The four transformations take the form:

$y = kf(x) \qquad y = f(x) + a \qquad y = f(kx) \qquad y = f(x + a)$

Each of these transformations is shown on the function $f(x) = x^3$.

$y = kf(x)$
Under this transformation the graph of the function is stretched (or shrunk if $k < 1$) along the y-axis.

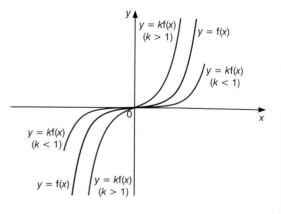

$y = f(x) + a$

Under this transformation the function is translated along the y-axis. If $a > 0$ then the graph of the function moves up (positive direction) and if $a < 0$ then the graph of the function moves down (negative direction).

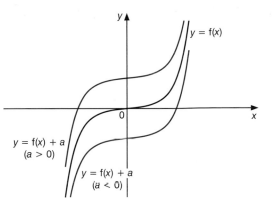

$y = f(kx)$

Under this transformation the graph of the function is shrunk (or stretched if $k < 1$) along the x-axis.

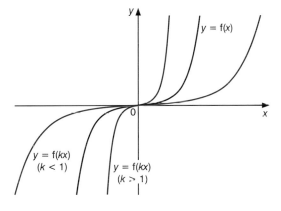

$y = f(x + a)$

Under this transformation the function is translated along the x-axis. If $a > 0$ then the graph of the function moves to the left (negative direction) and if $a < 0$ then the graph of the function moves to the right (positive direction).

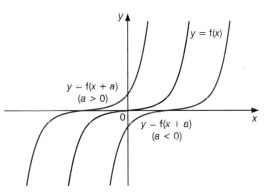

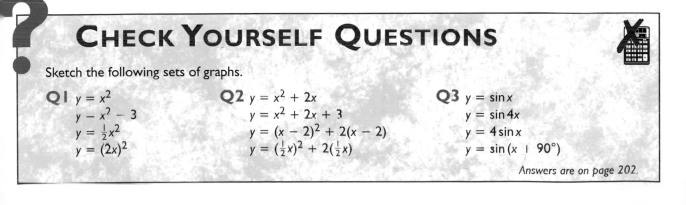

CHECK YOURSELF QUESTIONS

Sketch the following sets of graphs.

Q1 $y = x^2$
$y = x^2 - 3$
$y = \frac{1}{2}x^2$
$y = (2x)^2$

Q2 $y = x^2 + 2x$
$y = x^2 + 2x + 3$
$y = (x - 2)^2 + 2(x - 2)$
$y = (\frac{1}{2}x)^2 + 2(\frac{1}{2}x)$

Q3 $y = \sin x$
$y = \sin 4x$
$y = 4\sin x$
$y = \sin(x + 90°)$

Answers are on page 202.

UNIT 3: SHAPE, SPACE AND MEASURES

Geometric terms

≫ What do I need to know?

For your study of shape and space, you need to know and understand the following useful definitions.

≫ Angles and lines

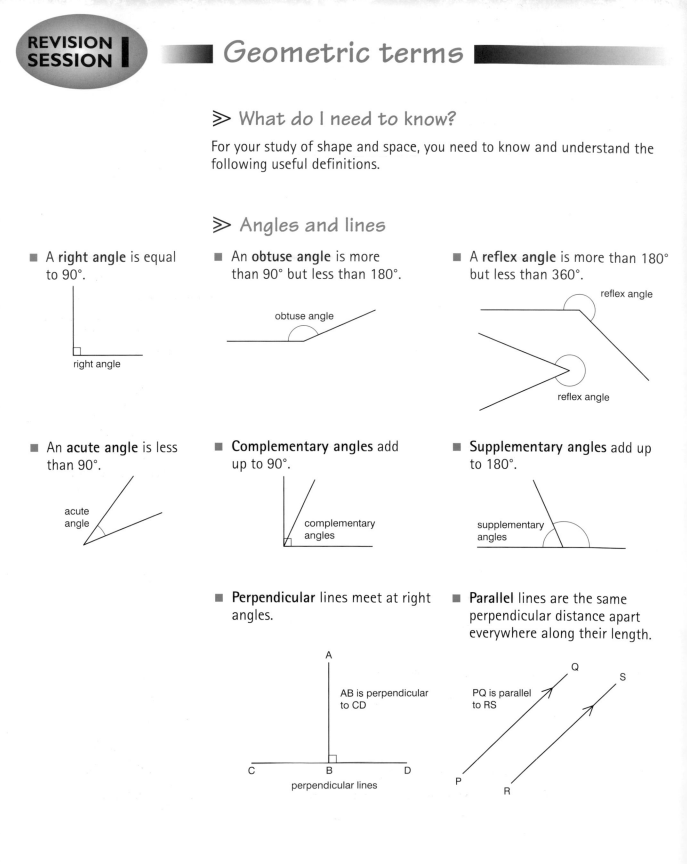

- A **right angle** is equal to 90°.

right angle

- An **obtuse angle** is more than 90° but less than 180°.

obtuse angle

- A **reflex angle** is more than 180° but less than 360°.

reflex angle

reflex angle

- An **acute angle** is less than 90°.

acute angle

- **Complementary angles** add up to 90°.

complementary angles

- **Supplementary angles** add up to 180°.

supplementary angles

- **Perpendicular** lines meet at right angles.

A

AB is perpendicular to CD

C   B   D

perpendicular lines

- **Parallel** lines are the same perpendicular distance apart everywhere along their length.

Q   S

PQ is parallel to RS

P

R

≫ Triangles

PROPERTIES OF TRIANGLES
The sum of the **interior angles** of a triangle is 180°.

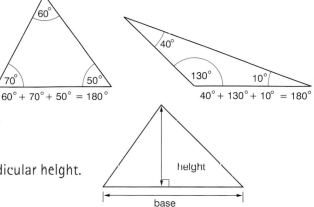

$$60° + 70° + 50° = 180°$$

$$40° + 130° + 10° = 180°$$

The smallest angle is always opposite the smallest side.

The largest angle is always opposite the largest side.

In general, the **area of a triangle** $= \frac{1}{2} \times$ base $\times$ perpendicular height.

TYPES OF TRIANGLE
You need to know about the following triangles and their properties.

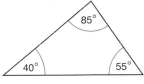

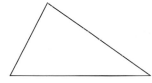

- An **acute-angled triangle** has all of its angles less than 90°.

- An **obtuse-angled triangle** has one of its angles greater than 90°.

- A **scalene triangle** has all its sides different and all its angles different.

- An **isosceles triangle** has two sides the same length and two angles the same size.

- An **equilateral triangle** has all three sides the same length and all three angles the same size.

- A **right-angled triangle** has one of its angles equal to 90°.

Worked example
Find the areas of the triangles.

a) Area of a triangle $= \frac{1}{2} \times$ base $\times$ perpendicular height

$\qquad = \frac{1}{2} \times 3 \times 5 \quad$ Perpendicular height is 5 cm.

$\qquad = 7.5 \, \text{cm}^2 \qquad$ Remember to include the units of area.

(a)

5 cm

3 cm

b) Area of a triangle $= \frac{1}{2} \times$ base $\times$ perpendicular height

$\qquad = \frac{1}{2} \times 7 \times 5 \quad$ Perpendicular height is 5 cm again.

$\qquad = 17.5 \, \text{cm}^2$

(b)

5 cm

7 cm

c) Area of a triangle $= \frac{1}{2} \times$ base $\times$ perpendicular height

$\qquad = \frac{1}{2} \times 4 \times 5 \quad$ Perpendicular height is still 5 cm.

$\qquad = 10 \, \text{cm}^2$

(c)

5 cm

4 cm

SHAPE, SPACE AND MEASURES

≫ Quadrilaterals

PROPERTIES OF QUADRILATERALS
The sum of the interior angles of a
quadrilateral is 360°.

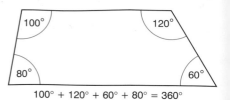

$100° + 120° + 60° + 80° = 360°$

TYPES OF QUADRILATERAL
You need to know about the following quadrilaterals and their properties.

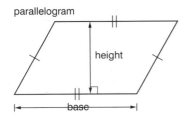

parallelogram

height

base

■ A **parallelogram** is a
quadrilateral in which the two
pairs of opposite sides are
equal and parallel.

The area of a parallelogram
= base × perpendicular height.

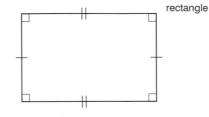

rectangle

■ A **rectangle** is a parallelogram
with four right angles.

The area of a rectangle
= base × height.

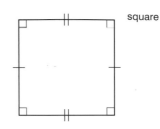

square

■ A **square** is a rectangle with
four equal sides.

The area of a square
= base × height
(where base = height).

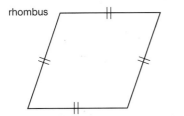

rhombus

■ A **rhombus** is a parallelogram
with four equal sides.

The area of a rhombus
= base × perpendicular height.

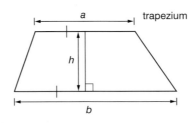

trapezium

a

h

b

■ A **trapezium** is a quadrilateral
with one pair of opposite sides
parallel.

The area of a trapezium

$= \frac{1}{2} \times \left(\begin{array}{c}\text{sum of parallel}\\\text{sides}\end{array}\right) \times \text{height}.$

This is usually written as

$\frac{1}{2}(a + b)h$

where a and b are the lengths
of the parallel sides and h is
the height.

kite

■ A **kite** is a quadrilateral with
two pairs of adjacent sides
equal.

Worked example

Find the areas of the quadrilaterals below.

a) Area of a parallelogram = base × perpendicular height

$\qquad = 8 \times 4$ Perpendicular height is 4 cm (not 5 cm).

$\qquad = 32 \text{ cm}^2$

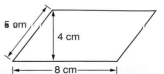

b) Area of trapezium = $\frac{1}{2}$ × (sum of parallel sides) × height

$\qquad = \frac{1}{2} \times (7 + 11) \times 6$ Perpendicular height is 6 cm.

$\qquad = \frac{1}{2} \times 18 \times 6$

$\qquad = 54 \text{ cm}^2$

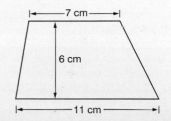

c) To find the area of a kite you can split the shape into two triangles and use:

area of a triangle = $\frac{1}{2}$ × base × perpendicular height

Area of kite = $\frac{1}{2} \times 6 \times 6.5 + \frac{1}{2} \times 6 \times 8.5$ Height of top triangle
$\qquad\qquad\qquad\qquad\qquad\qquad\qquad\qquad = 15 - 8.5 = 6.5 \text{ cm}.$

$\qquad = 19.5 + 25.5$

$\qquad = 45 \text{ cm}^2$

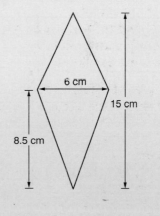

An easier way to find the area of a kite is to multiply the width by the height and divide by 2.

Area of kite $= \frac{1}{2}$ × width × height

≫ Polygons

Any shape enclosed by straight lines is called a **polygon**. Polygons are named according to their number of sides.

- A **regular polygon** has all sides equal and all angles the same.

- A **convex polygon** has no interior angle greater than 180°.

- A **concave** (or **re-entrant**) **polygon** has at least one interior angle greater than 180°.

Number of sides	Name of polygon
3	triangle
4	quadrilateral
5	pentagon
6	hexagon
7	heptagon
8	octagon
9	nonagon
10	decagon

SHAPE, SPACE AND MEASURES

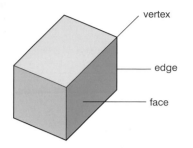
vertex

edge

face

≫ Solids

A **solid** is a three-dimensional shape such as cube, cuboid, prism, cylinder, sphere, pyramid or cone.

You need to understand the following definitions relating to solids.

- A **face** is the surface of a solid which is enclosed by edges.

- An **edge** is a straight line where two faces meet.

- A **vertex** is the point where three or more edges meet.

TYPES OF SOLID

You should be aware of the following solids and their properties.

- A **cube** is a three-dimensional shape with six square faces.

- A **cuboid** is a three-dimensional shape with six rectangular faces. Opposite faces are equal in size.

- A **prism** is a three-dimensional shape with uniform cross-section. The prism is usually named after the shape of the cross-sectional area.

triangular prism

hexagonal prism

octagonal prism

Cubes and cuboids are also prisms as they have uniform cross-sections.

- A **cylinder** is a prism with a uniform circular cross-section.

- A **sphere** is a three-dimensional shape in which the surface is always the same distance from the centre.

sphere   hemisphere

- A **hemisphere** is one half of a sphere.

- A **pyramid** is a three-dimensional shape with a polygon-shaped base and the remaining triangular faces meeting at a vertex. The pyramid is usually named after the shape of the polygon forming the base.

triangular pyramid

square-based pyramid

hexagon-based pyramid

cone

- A **cone** is a pyramid with a circular base.

≫ Nets

A net is a pattern that can be cut out and folded to form a 3D shape.
The following nets are the most common.

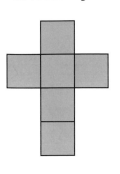

cube

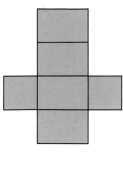

triangular prism

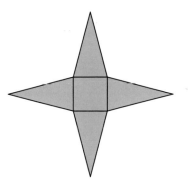

cuboid

square-based pyramid

CHECK YOURSELF QUESTIONS

Q1 Find the value of each angle that is marked with a letter.

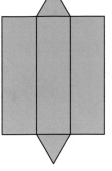

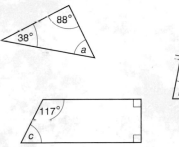

Q2 Write down the value of the smallest angle.

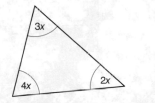

Q3 Find the areas of the following polygons.

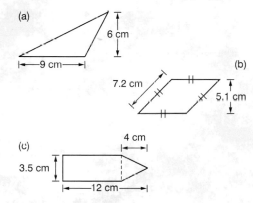

Q4 Draw nets of the following solids.

a) tetrahedron (a pyramid in which all the faces are equilateral triangles)

b) hexagonal prism

Answers are on page 203.

Constructions

≫ What will I need?

You will need the following equipment to construct diagrams, produce scale drawings and find the locus of points.

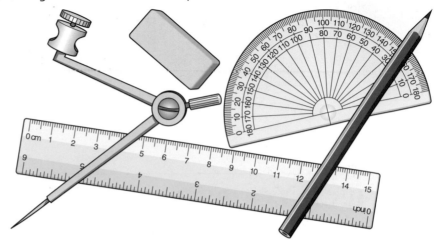

- a sharp pencil and an eraser
- a ruler
- a pair of compasses
- a protractor

You should always make your constructions as accurate as you possibly can – the examiner expects measurements of length to be accurate to the nearest millimetre and measurements of angles to be correct to the nearest degree.

≫ Constructing triangles

When working with a triangle ABC, it is conventional to represent the side opposite angle A by the letter a, the side opposite angle B by the letter b and the side opposite angle C by the letter c, as shown here.

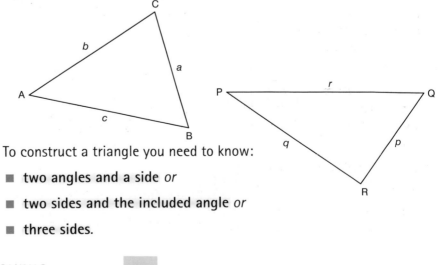

To construct a triangle you need to know:

- two angles and a side *or*
- two sides and the included angle *or*
- three sides.

≫ Given two angles and a side

Worked example
Construct triangle ABC in which $\angle A = 48°$, $\angle B = 76°$ and $b = 6$ cm.

1 Draw a rough sketch.

2 Draw the line segment AC = 6 cm.

3 Draw an angle of 56° at C.

4 Draw an angle of 48° at A.

5 Label the point of intersection as B.

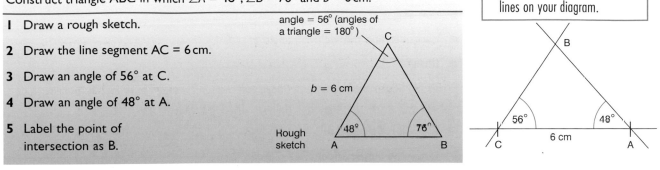

angle = 56° (angles of a triangle = 180°)

Rough sketch

≫ Given two sides and the included angle

Worked example
Construct triangle PQR in which $p = 8$ cm, $q = 6.5$ cm and $\angle QRP = 53°$.

1 Draw a rough sketch.

2 Draw the line segment QR = 8 cm.

3 Draw an angle of 53° at R.

4 Construct the point P which is 6.5 cm from R, using your compasses with centre R and radius 6.5 cm.

5 Join QP.

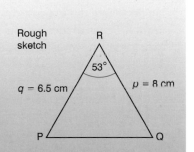

Rough sketch

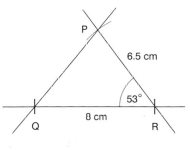

≫ Given three sides

Worked example
Construct triangle XYZ in which $x = 5.8$ cm, $y = 8.2$ cm and $z = 5.5$ cm.

1 Draw a rough sketch.

2 Draw the line segment XZ = 8.2 cm (the longest side).

3 With centre Z and radius 5.8 cm, use your compasses to draw a circular arc giving points 5.8 cm from Z.

4 With centre X and radius 5.5 cm, use your compasses to draw a circular arc giving points 5.5 cm from X.

5 Label the point of intersection of the two arcs as Y.

6 Join ZY and XY.

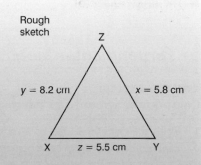

Rough sketch

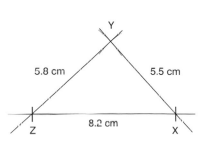

Worked example

Construct triangle ABC in which $a = 7.6$ cm, $b = 5.6$ cm and $\angle ABC = 44°$.

Before constructing the triangle, remind yourself what you need to know:

- two angles and a side *or*
- two sides and the included angle *or*
- three sides.

In this particular case you *do not have sufficient information to draw the triangle*, since you are not given the included angle, so proceed as follows.

1 Draw a rough sketch.

2 Draw the line segment BC = 7.6 cm.

3 Draw an angle of 44° at B.

4 With compasses set to a radius of 5.6 cm, and centred on C, draw a circular arc giving points 5.6 cm from C.

5 You will find that there are two points where this arc cuts the line from B. Either of these points will give a point A which satisfies the conditions for the triangle.

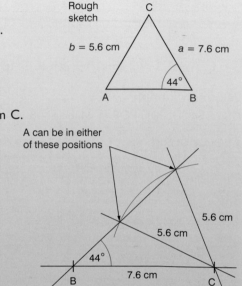

+ *HINT*

▸ Because there are two possible solutions to the triangle, this is sometimes called the ambiguous case.

≫ Geometrical constructions

You should be able to carry out the following constructions, using only a ruler and a pair of compasses:

- the **perpendicular bisector of a line**
- the **perpendicular from a point on a straight line**
- the **perpendicular from a point to a straight line**
- the **angle bisector**.

≫ Perpendicular bisector of a line

Worked example

Construct the perpendicular bisector of a line AB.

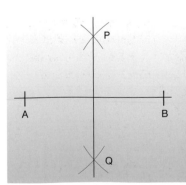

1 With the compasses set to a radius greater than half the length of AB, and centred on A, draw arcs above and below the line.

2 With the compasses still set to the same radius, and centred on B, draw arcs above and below the line, so that they cut the first arcs.

3 Join the points where these arcs cross (P and Q). This line is the perpendicular bisector of AB.

≫ Perpendicular from a point on a straight line

Worked example

Construct the perpendicular at the point X on a line.

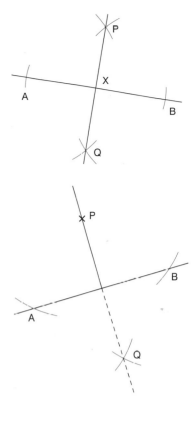

1   With the compasses set to a radius of about 5 cm, and centred on X, draw arcs to cut the line at A and B.

2   Now construct the perpendicular bisector of the line segment AB, as before.

≫ Perpendicular from a point to a straight line

Worked example

Construct the perpendicular to the line from a point P above the line.

1   With the compasses set to a suitable radius, and centred on P, draw arcs to cut the line at A and B.

2   With the compasses set to a radius greater than half the length of AB, and centred on A, draw an arc on the opposite side of the line from P.

3   With the compasses still set to the same radius, and centred on B, draw an arc to cut the arc drawn in step 2, at Q.

4   Join PO.

≫ The angle bisector

Worked example

Construct the bisector of an angle ABC.

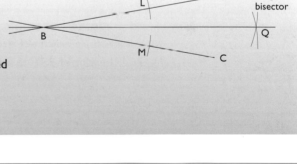

1   With the compasses set to a radius of about 5 cm, and centred on B, draw arcs to cut BA at L and BC at M.

2   With the compasses set to the same radius, and centred on L, draw an arc between BA and BC.

3   With the compasses still set to the same radius, and centred on M, draw an arc to cut the arc between BA and BC.

4   Label the point where the arcs cross as Q.

5   Join BQ. This is the bisector of the angle ABC.

? CHECK YOURSELF QUESTIONS

Q1 Draw any triangle ABC, then construct the perpendicular bisectors of the two longest sides. Label the point of intersection of these perpendicular bisectors O. Draw the circle centre O and radius OA.

Q2 Construct triangle PQR where p = 4.5 cm, q = 6 cm and r = 7.5 cm. What do you notice about your triangle?

Q3 Draw the line segment PQ so that PQ = 8 cm. Construct the perpendicular bisector of PQ. Construct the points R and S so that they lie on the perpendicular bisector a distance of 3 cm from the line PQ.
Join PR, RQ, QS and SP.
What is the special name given to this quadrilateral?

Answers are on page 203.

Maps and scale drawings

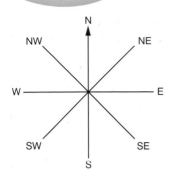

≫ What are bearings and scale drawings?

Bearings are a useful way of describing directions. They are usually given as three-figure numbers, so a bearing of 50° would usually be written as 050°.

Scale drawings are drawings or diagrams that are scaled according to some rule. The scale for the drawing or diagram must be clearly stated.

≫ Using bearings

Worked example

The bearing of a ship from a lighthouse is 035°. What is the bearing of the lighthouse from the ship?

By drawing a sketch of the situation, you can see more clearly that the required bearing is 215°.

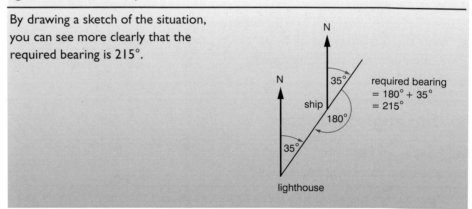

required bearing
= 180° + 35°
= 215°

Worked example

A boat sails due east from a point A for 12 km to a position B, then for 6 km on a bearing of 160° to a point C. Using a scale of 1 cm to represent 2 km, show this on a scale drawing and use this to find the bearing and distance from A to C.

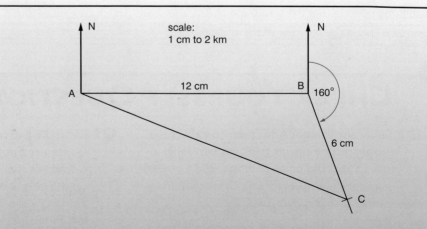

The bearing from A to C is measured as 112° and the distance from A to C is 7.6 cm which represents 15.2 km (remembering to use the given scale to convert back to kilometres).

? CHECK YOURSELF QUESTIONS

Q1 Find the bearing of B from A if the bearing of A from B is:

a) 090°

b) 120°

c) 345°

d) 203°.

Q2 A plane takes off heading north-west and is told to take a left-hand turn at 5000 feet. What bearing is the plane now headed on?

Q3 On a map, the scale is given as 1 inch to 5 miles.

a) What distance is represented by $4\frac{1}{2}$ inches on the map?

b) A road is 36 miles long. How long is this on the map?

Q4 Two explorers set off from the same point one morning. One explorer travels at 4.5 mph on a bearing of 036° and the other explorer travels at 5.5 mph on a bearing of 063°. Using a scale of 0.5 cm to 1 mile, calculate how far they are apart after 2 hours.

Answers are on page 204.

Imperial and metric units

≫ Why are there two?

Until fairly recently, the UK used the imperial system for measuring. Although metric units (based on a decimal system) are replacing the imperial system, you should be familiar with both.

≫ Imperial measure

Length

12 inches (in)	= 1 foot (ft)
3 feet	= 1 yard (yd)
1760 yards	= 1 mile

Capacity

20 fluid ounces (fl oz)	= 1 pint (pt)
8 pints	= 1 gallon (gall)

Weight

16 ounces (oz)	= 1 pound (lb)
14 pounds	= 1 stone (st)
8 stones	= 1 hundredweight (cwt)
20 hundredweights	= 1 ton

≫ Metric (SI) measure

Length

10 millimetres (mm)	= 1 centimetre (cm)
100 centimetres	= 1 metre (m)
1000 millimetres	= 1 metre
1000 metres	= 1 kilometre (km)

Capacity

10 millilitres (ml)	= 1 centilitre (cl)
1000 millilitres	= 1 litre

Weight

1000 milligrams	= 1 gram
1000 grams	= 1 kilogram (kg)
1000 kilograms	= 1 tonne (t)

≫ Conversion factors

You will need to know these approximate conversions which may be tested in the examination.

Imperial	Metric
1 inch	2.5 centimetres
1 foot	30 centimetres
5 miles	8 kilometres
1 litre	1.75 pints
1 gallon	4.5 litres
2.2 pounds	1 kilogram

NOTATION

When working with approximations, you may see the notation:

$1 \text{ inch} \approx 2.5 \text{ cm}$

which is a short way of writing '1 inch is approximately equal to 2.5 cm'.

+ HINT

▸ The symbol $\approx$ means 'is approximately equal to'.

Worked example

How many millilitres are there in one pint?

1 gallon ≈ 4.5 litres	
8 pints ≈ 4.5 litres	As 1 gallon = 8 pints
8 pints ≈ 4500 millilitres	As 1 litre = 1000 millilitres
1 pint ≈ 4500 ÷ 8	Dividing by 8 to find 1 pint.
= 562.5 millilitres	

So there are 560 millilitres (to the nearest 10 ml) in one pint.

+ HINT

▸ It is not reasonable to present the answer too accurately as the conversions are only approximate.

? CHECK YOURSELF QUESTIONS

Q1 How many yards are there in one kilometre?

Q2 How many metres are there in 250 yards?

Q3 How many kilograms are there in one stone?

Answers are on page 204.

▰ Locus of points

≫ What is a locus?

A **locus** is the path followed by points that satisfy some given rule.

≫ Some common loci

Worked example

Find the locus of a point moving so that it is a fixed distance from a point O.

The locus of a point moving so that it is a fixed distance from a point O is a circle with centre O.

locus of points at fixed distance from O

•O

Worked example

Find the locus of a point moving so that it is always equidistant from two fixed points A and B.

The locus of a point moving so that it is equidistant from two fixed points A and B is the perpendicular bisector of the line AB.

locus of points at equal distance from A and B

A ——————— B

Worked example

Find the locus of a point moving so that it is a fixed distance from the line PQ.

The locus of a point moving so that it is a fixed distance from the line PQ is a pair of lines parallel to PQ, along with the semicircles at the points P and Q.

locus of points at a fixed distance from PQ

P ——————— Q

Worked example

Find the locus of a point moving so that it is a fixed distance from two lines AB and CD.

The locus of a point moving so that it is a fixed distance from two lines AB and CD is the pair of bisectors of the angles between the two lines (drawn from the point where the lines cross).

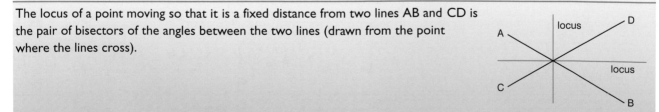

A   locus   D

locus

C   B

Worked example

A tree is situated 4 m away from a wall. Draw a scale diagram and identify the points which are 3 metres away from the wall and $2\frac{1}{2}$ m away from the tree.

The points P and Q are 3 m from the wall and $2\frac{1}{2}$ m from the tree.

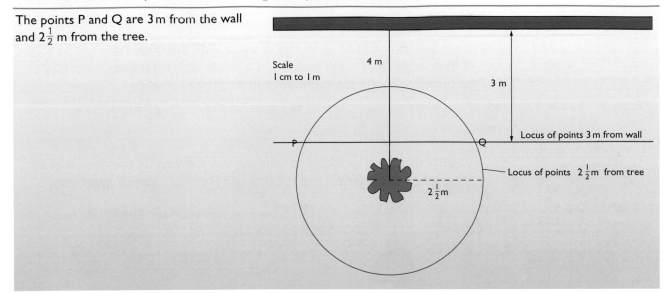

Scale
1 cm to 1 m

4 m

3 m

P · · Q

Locus of points 3 m from wall

Locus of points $2\frac{1}{2}$ m from tree

$2\frac{1}{2}$ m

CHECK YOURSELF QUESTIONS

Q1 Find the locus of points that are 3 cm from the line AB, which is of length 5 cm.

Q2 Find the locus of points that are equidistant from two points P and Q which are 3 cm apart.

Q3 XY is a fixed line of length 8 cm and P is a variable point. The distance of P from X is 6.2 cm and the distance of P from Y is 3.7 cm. Show all of the possible positions of P on a diagram.

Q4 XYZ is an equilateral triangle of sides 5 cm. Show all of the points that are less than 2 cm from the edges of the triangle.

Answers are on page 205.

Symmetry

≫ What is symmetry?

There are two types of symmetry. A shape with **line symmetry** can be folded in half so that both sides match and a shape with **rotational symmetry** can be turned so that it looks the same in a different position.

≫ Line symmetry

When a shape can be folded so that one half fits exactly over the other half, the shape is **symmetrical** and the fold line is called a **line of symmetry**.

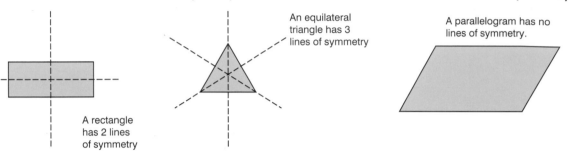

An equilateral triangle has 3 lines of symmetry

A rectangle has 2 lines of symmetry

A parallelogram has no lines of symmetry.

Tracing paper is useful when you need to identify lines of symmetry or to draw reflections. You can use tracing paper in the mathematics examination.

≫ Plane of symmetry

A **plane of symmetry** divides a solid into two equal halves.

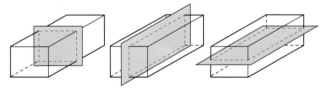

A cuboid has 3 planes of symmetry

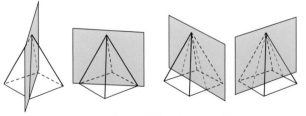

A square-based pyramid has 4 planes of symmetry

≫ Rotational symmetry

When a shape can be rotated about its centre to a new position, so that it fits exactly over its original position, then the shape has **rotational symmetry**.

The number of different positions tells you the **order** of rotational symmetry. An equilateral triangle has rotational symmetry of order 3.

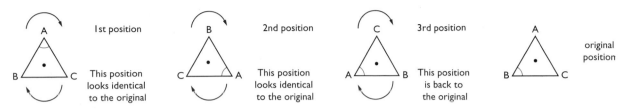

Once again, tracing paper is useful when you need to identify rotational symmetry. You can use tracing paper in the mathematics examination.

?

CHECK YOURSELF QUESTIONS

Q1 Write down all of the letters below that have:

a) a horizontal line of symmetry

b) a vertical line of symmetry

c) both horizontal and vertical lines of symmetry

d) no line symmetry

e) rotational symmetry of order 2.

Q2 How many lines of symmetry does a 50p piece have?

Q3 How many planes of symmetry do the following solids have?

a) triangular prism

b) hexagonal prism

ABCDEFGHIJK

Answers are on page 205.

Transformations

≫ What is a transformation?

If a point or a collection of points is moved from one position to another it undergoes a **transformation**. The **object** is the point or collection of points before the transformation and the **image** is the point or collection of points after the transformation.

The transformations you will need to know about include:

- reflection
- rotation
- enlargement
- translation.

HINT

▶ Take care with non-vertical or non-horizontal lines — tracing paper is useful in these questions.

HINT

▶ Subsequent transformations on ABC... are usually written as A'B'C'..., A"B"C"... or else $A_1B_1C_1...$, $A_2B_2C_2...$ etc.

≫ Reflection

A reflection is a transformation in which any two corresponding points on the object and image are the same distance away from a fixed line (called the **line of symmetry** or **mirror line**).

You can define a reflection by giving the position of the line of symmetry.

The image of ABCDE is labelled A'B'C'D'E' and corresponding pairs of points in the image and the object are **equidistant** from the line of symmetry.

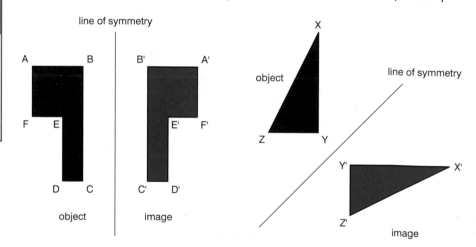

≫ Rotation

A rotation is a transformation in which lines from any two corresponding points on the object and image make the same angle at a fixed point (called the **centre of rotation**).

You define a rotation by giving the position of the centre of rotation, along with the angle and direction of the rotation.

The diagram shows a rotation through 90° in an anticlockwise direction (called a rotation of $^+$90°) about the centre of rotation O.

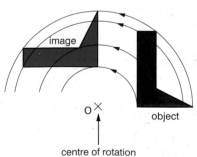

To find the centre of rotation you should join corresponding points on the object and image with straight lines and draw the **perpendicular bisectors** of these lines. The centre of rotation lies on the intersection of these bisectors.

To find the angle of rotation you should join corresponding points on the object and image to the centre of rotation. The angle between these lines is the angle of rotation.

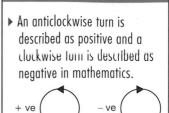

HINT

▸ An anticlockwise turn is described as positive and a clockwise turn is described as negative in mathematics.

+ ve   − ve

≫ Enlargement

An enlargement is a transformation in which the distance between a point on the image and a fixed point (called the **centre of enlargement**) is a factor of the distance between the corresponding point on the object and the fixed point.

You can define an enlargement by giving the position of the centre of enlargement along with the factor (called the **scale factor**).

This diagram on the right shows an enlargement, scale factor 3, based on the centre of enlargement O.

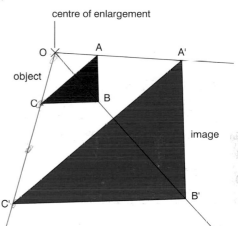

OA = 1.1 cm   OA′ = 3 × 1.1 = 3.3 cm
OB = 1.7 cm   OB′ = 3 × 1.7 = 5.1 cm
OC = 1.35 cm   OC′ = 3 × 1.35 = 4.15 cm

To find the centre of enlargement you join corresponding points – on the object and image – with straight lines. The centre of enlargement lies at the intersection of these straight lines.

You can find the scale factor (SF) of an enlargement as follows:

$$SF = \frac{\text{distance from the centre of a point on the image}}{\text{distance from the centre of the corresponding point on the object}}$$

or $$SF = \frac{\text{distance between two points on the image}}{\text{distance between two corresponding points on the object}}$$

A scale factor greater than 1 will enlarge the object.

A scale factor less than 1 will have the effect of reducing (or shrinking) the object.

Worked example

The points A(3, 8), B(7, 8), C(7, ⁻4) and D(3, ⁻2) are joined to form a trapezium which is enlarged, scale factor $\frac{1}{2}$, with (⁻5, ⁻6) as the centre of enlargement.
Draw ABCD on a graph and hence find A′B′C′D′.

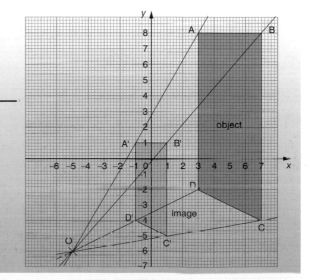

The solution is shown on the right.

OA = 16.12   OA′ = $\frac{1}{2}$ × 16.12 = 8.06

OB = 18.44   OB′ = $\frac{1}{2}$ × 18.44 = 9.22

OC = 12.2   OC′ = $\frac{1}{2}$ × 12.2 = 6.1

OD = 8.94   OD′ = $\frac{1}{2}$ × 8.94 = 4.57

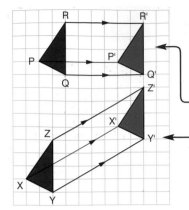

≫ Translation

A translation is a transformation in which the distance and direction between any two corresponding points on the object and image are the same. You can define a translation by giving the distance and direction of the translation.

This translation can be described as a movement of five units to the left.

This translation can be described as a movement of seven units to the right and four units upwards.

You can use **vector notation** to write a movement of seven units to the right and four units up as $\binom{7}{4}$. In general, you can write:

$$\binom{a}{b} \Rightarrow \binom{\text{number of units to the right in the positive } x\text{-direction}}{\text{number of units upwards in the positive } y\text{-direction}}$$

Worked example

The triangle ABC with coordinates $(1, 1), (3, 2)$ and $(2, 5)$ undergoes a translation of $\binom{2}{-6}$ to A′B′C′. Sketch the triangles ABC and A′B′C′ and write down the translation which will return A′B′C′ to ABC.

The vector $\binom{2}{-6}$ describes a movement of 2 units to the right and ⁻6 units upwards (i.e. 6 units downwards).

The translation which will return A′B′C′ to ABC is $\binom{-2}{6}$.

≫ Combinations of transformations

Combinations of the same type of transformation or combinations of different types of transformation can usually be described as a single transformation.

Worked example

If R is a reflection in the y-axis and T is a rotation about the origin of ⁻90°, show (on separate diagrams) the image of the triangle XYZ with vertices X(2, 1), Y(2, 5) and Z(4, 2), under the combined transformations:

a) R followed by T   **b)** T followed by R.

Which single transformation will return each of these combined transformations back to their original position?

a) The single transformation that will return A″B″C″ to ABC is a reflection in the line $y = {}^-x$.

b) The single transformation that will return A″B″C″ to ABC is a reflection in the line $y = x$.

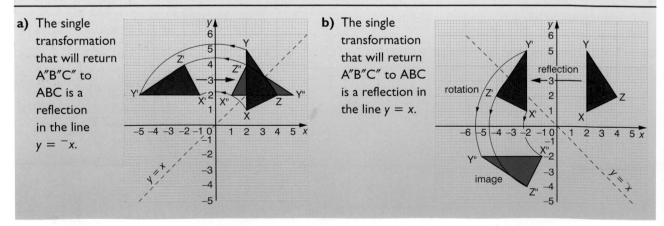

CHECK YOURSELF QUESTIONS

Q1 The triangle ABC with coordinates A(3, 1), B(4, 1) and C(4, 4) is reflected in the line $y = x$ then the image undergoes a reflection in the line $x = 0$. The resulting triangle is then rotated through $^-90°$ about $(^-1, 4)$ to A'B'C'. What single transformation would map ABC onto A'B'C'?

Q2 Find the images of the point (3, 4) reflected in the following lines.

a) $x = 0$ **b)** $y = 0$

c) $x = 2$ **d)** $y = x$

Q3 Draw the images of the following objects after:

i)   a rotation of $^+90°$ about the given centre of rotation O

ii) a rotation of $^-90°$ about the given centre of rotation O.

a)   **b)**   **c)**

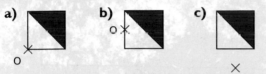

Q4 Draw the images of the following objects after an enlargement, scale factor $\frac{3}{2}$, with centre of enlargement O.

a)   **b)**

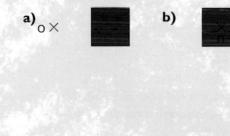

Q5 Draw the images of △ABC (triangle ABC) after the following translations.

a) $\begin{pmatrix} 3 \\ 2 \end{pmatrix}$ **b)** $\begin{pmatrix} ^-4 \\ 0 \end{pmatrix}$ **c)** $\begin{pmatrix} ^-2 \\ ^-5 \end{pmatrix}$

Write down the translation which will return the image to ABC in each case.

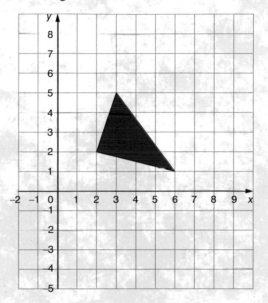

Q6 Find the centre of rotation and the angle of rotation for the following transformation.

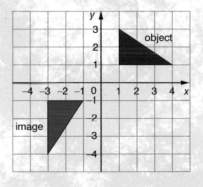

Answers are on page 206.

Angle properties

≫ What *do* I need to know?

You need to know the following angle properties, and use them when explaining work with lines and angles. You also need to remember that:

- the **interior angles of a triangle add up to 180°** and the **interior angles of a quadrilateral add up to 360°**

- the **exterior angle plus the adjacent interior angle add up to 180°** (angles on a straight line) and the **sum of all the exterior angles of a polygon is 360°**.

≫ Properties of angles

- Angles at a point add up to 180°.

  $90° + 90° = 180°$

  $a + b = 180°$

- Angles at a point add up to 360°.

  $90° + 90° + 90° + 90° = 360°$

  $v + w + x + y + z = 360°$

- When two straight lines intersect the (vertically) opposite angles are equal.

  $180° - a$ a a $180° - a$

≫ Angles between parallel lines

- A **transversal** is a line which cuts two or more parallel lines.

  transversal

- **Corresponding** angles are equal.

  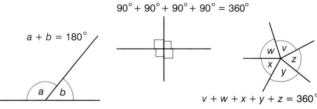

  $a = e$
  $b = f$
  $c = g$
  $d = h$
  } All pairs of corresponding angles

- **Alternate angles** (or Z angles) are equal.

  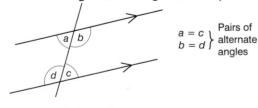

  $a = c$
  $b = d$
  } Pairs of alternate angles

- **Interior angles** add up to 180° (i.e. interior angles are **supplementary**).

  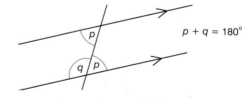

  $p + q = 180°$

Worked example

Find the values of a, b and c in this diagram. Give reasons for your answers.

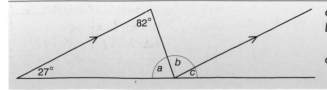

$a = 71°$ The angles of the triangle add up to 180°.
$b = 82°$ b is an alternate angle between the two given parallel lines.
$c = 27°$ The angles on a straight line add up to 180°.

≫ Angle sum of a triangle

You can show that the angle sum of a triangle is 180°, as follows.

Let the angles of the triangle ABC be α, β and γ.

Draw a line segment CD, parallel to AB.

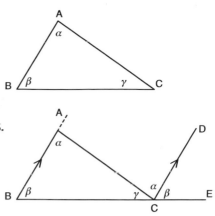

Now $\angle ACD = \angle BAC = \alpha$ Alternate angles between parallel lines.

      $\angle ABC = \angle DCE = \beta$ Corresponding angles between parallel lines.

      BC is a straight line, so $\alpha + \beta + \gamma = 180°$
                As angles on a straight line add up to 180°.

So the angle sum of a triangle is 180°.

≫ Angle sum of a polygon

You can find the sum of the angles of a polygon by dividing the polygon into a series of triangles where each triangle has an angle sum of 180°.

■ A four-sided polygon can be split into two triangles.

Angle sum = 2 × 180° = 360°

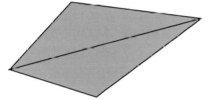

■ A six-sided polygon can be split into four triangles.

Angle sum = 4 × 180° = 720°

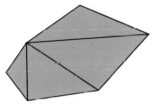

■ A seven-sided polygon can be split into five triangles.

Angle sum = 5 × 180° = 900°

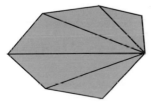

From the above, you can see that an n-sided polygon can be split into $(n - 2)$ triangles.

Angle sum of an n-sided polygon = $(n - 2) \times 180°$

> **➕ HINT**
>
> ▸ The angle sum of an n-sided polygon is sometimes presented as $(2n - 4)$ right angles. This formula gives exactly the same results as $(n - 2) \times 180°$.

≫ Exterior angles

Another useful result involves the exterior angles. Since the sum of the exterior angles is 360°, each exterior angle of a regular polygon is 360° ÷ *n*.

You can use this fact to find the interior angle of a regular polygon.

Interior angle = 180° − exterior angle

The angle sum of an *n*-sided regular polygon can be found in the same way as the angle sum of a non-regular polygon.

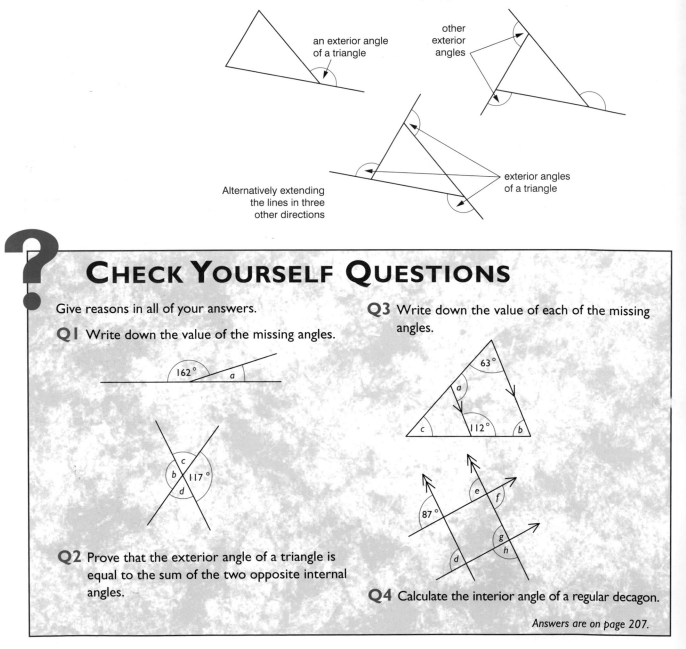

an exterior angle of a triangle

other exterior angles

exterior angles of a triangle

Alternatively extending the lines in three other directions

CHECK YOURSELF QUESTIONS

Give reasons in all of your answers.

Q1 Write down the value of the missing angles.

Q2 Prove that the exterior angle of a triangle is equal to the sum of the two opposite internal angles.

Q3 Write down the value of each of the missing angles.

Q4 Calculate the interior angle of a regular decagon.

Answers are on page 207.

≫ What will I need to know?

You need to understand the terms **congruent** and **similar** for the examination.

≫ Congruent triangles

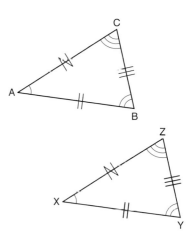

Two triangles are congruent if one of the triangles can be fitted exactly over the other triangle so that all corresponding angles are equal and all corresponding sides are equal.

AB = XY   BC = YZ   CA = ZX
∠A = ∠X   ∠B = ∠Y   ∠C = ∠Z

To prove that two triangles are congruent, it is not necessary to prove that all of the above conditions are true. The following minimum conditions are sufficient to show that two triangles are congruent.

Two triangles are congruent if:

■ **two angles and a side** of one triangle are equal to two angles and the corresponding side of the other (**AAS**)

■ **two sides and the included angle** of one triangle are equal to two sides and the included angle of the other (**SAS**)

■ the **three sides** of one triangle are equal to the three sides of the other (**SSS**).

If the triangles are right-angled then the two triangles are congruent if:

■ the **hypotenuses and one other side** are equal on both triangles (**RHS**).

> **＋ HINT**
>
> ▸ If the two triangles ABC and XYZ are congruent then you can write △ABC ≡ △XYZ.
> ▸ The symbol ≡ is a short way of writing 'is congruent to'.

Worked example

Look at the diagram and then write down a pair of triangles that are congruent.
Explain why they are congruent.

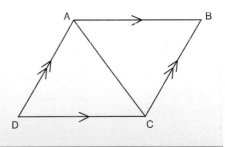

△ABC ≡ △CDA

∠BAC = ∠DCA	Alternate angles between parallel lines AB and DC.
∠DAC = ∠BCA	Alternate angles between parallel lines AD and BC .
AC = CA	Common line.
so △ABC ≡ △CDA	As two angles and a side of one triangle are equal to two angles and the corresponding side of the other (AAS).

Two angles and a side of one triangle are equal to two angles and the corresponding side of the other (AAS).

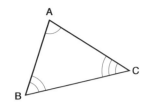

≫ Similar triangles

Two triangles are similar if one of the triangles is an enlargement of the other triangle so that all corresponding angles are equal and corresponding sides are in the same ratio.

$$\angle A = \angle X$$
$$\angle B = \angle Y$$
$$\angle C = \angle Z$$

If two triangles are similar then the ratios of the corresponding sides are equal.

$$\frac{AB}{XY} = \frac{BC}{YZ} = \frac{CA}{ZX} \quad \text{or} \quad \frac{XY}{AB} = \frac{YZ}{BC} = \frac{ZX}{CA}$$

To prove that two triangles are similar, it is not necessary to prove that all of the above conditions are true. The following minimum conditions are sufficient to show that two triangles are similar.

Two triangles are similar if:
- **two angles** of one triangle are equal to two angles of the other
- **two pairs of sides** are in the same ratio and the **included angles are equal**
- **three pairs of sides** are in the same ratio.

Worked example

Triangles ABC and PQR are similar.

Calculate: **a)** $\angle PQR$ **b)** QR   **c)** AC.

What can you say about the ratio of the lengths of corresponding sides of the two triangles?
What can you say about the ratio of the areas of the two triangles?

(Triangle ABC: AB = 4 cm, BC = 3 cm, right angle at B)
(Triangle PQR: PQ = 5 cm, PR = 6.25 cm)

a) $\angle PQR = \angle ABC = 90°$ As the triangles are similar then the corresponding angles are equal.

b) In order to find QR, you need to know the ratio of corresponding sides.

$$AB : PQ = 4 : 5 \text{ or } \frac{AB}{PQ} = \frac{4}{5}$$

$$\frac{BC}{QR} = \frac{AB}{PQ}$$

so $\quad \dfrac{BC}{QR} = \dfrac{4}{5}$

$\dfrac{3}{QR} = \dfrac{4}{5}$ As BC = 3 cm.

$\dfrac{QR}{3} = \dfrac{5}{4}$ Turning both sides upside-down.

$QR = \dfrac{5}{4} \times 3$ Multiplying both sides by 3.

$QR = 3.75 \, cm$

c) In order to find AC, you need to use the ratio of corresponding sides.

$\dfrac{AB}{PQ} = \dfrac{4}{5}$ As before.

$\dfrac{AC}{PR} = \dfrac{AB}{PQ}$

so $\quad \dfrac{AC}{PR} = \dfrac{4}{5}$

$\dfrac{AC}{6.25} = \dfrac{4}{5}$ As PR = 6.25 cm.

$$AC = \frac{4}{5} \times 6.25 \quad \text{Multiplying both sides by 6.25.}$$

$$AC = 5\,cm$$

Using the fact that the area of a triangle $= \frac{1}{2} \times$ base $\times$ perpendicular height:

area of triangle ABC $= \frac{1}{2} \times 4 \times 3 = 6\,cm^2$

area of triangle PQR $= \frac{1}{2} \times 5 \times 3.75 = 9.375\,cm^2$

ratio of lengths $= 4 : 5$

ratio of areas $= 6 : 9.375$

$\qquad\qquad = 6000 : 9375$ Multiplying both sides of the ratio by 1000 to get whole numbers.

$\qquad\qquad = 16 : 25$ Cancelling both sides.

You should notice that the ratio of the areas is equal to the ratio of the lengths squared (i.e. $4^2 : 5^2$).

CHECK YOURSELF QUESTIONS

Q1 Show that $\triangle PQT \equiv \triangle SRT$ and hence find the length RS.

Q2 Find the length AC in the following diagram.

Q3 Find the length MN in the following diagram.

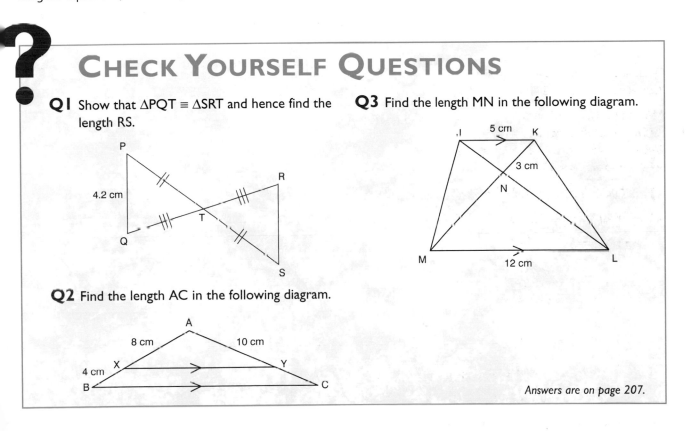

Answers are on page 207.

Pythagoras' theorem in two dimensions

≫ What is Pythagoras' theorem?

For any **right-angled triangle**, the square of the length of the hypotenuse is equal to the sum of the squares of the lengths of the other two sides.

$$a^2 + b^2 = c^2$$

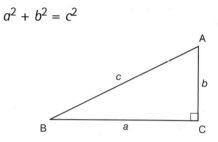

For a right-angled triangle, the side opposite the right angle is called the **hypotenuse** and this is always the longest side.

≫ Using Pythagoras' theorem

Worked example

Find the length of the hypotenuse in this triangle.

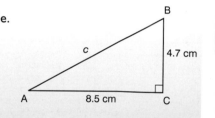

Using Pythagoras' theorem:

$$a^2 + b^2 = c^2$$

$$c^2 = a^2 + b^2$$

$c^2 = 4.7^2 + 8.5^2$ Substituting values for each of the given lengths.

$c^2 = 22.09 + 72.25$ Squaring individual lengths.

$c^2 = 94.34$

$c = 9.712\,878$ Taking square roots of both sides to find c.

$c = 9.71$ cm (3 s.f.)   Rounding to an appropriate degree of accuracy.

Worked example

Calculate the height of this isosceles triangle.
Leave your answer in surd form.

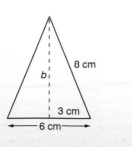

HINT

 means 'Do not use a calculator.'

The isosceles triangle can be split into
two right-angled triangles to find the height.

Using Pythagoras' theorem:

$a^2 + b^2 = c^2$

$3^2 + b^2 = 8^2$ Substituting values for each of the given lengths – in this case the length of the hypotenuse is known.

$9 + b^2 = 64$ Squaring individual lengths.

$b^2 = 64 - 9$ Making the height the subject.

$b^2 = 55$

$b = \sqrt{55}$ Taking square roots of both sides to find the height.

$b = \sqrt{55}$ cm   Leaving the answer in surd form.

CHECK YOURSELF QUESTIONS

Q1 Without using a calculator, find the length of a diagonal of a square of side 5 cm. Leave your answer in surd form.

Q2 Find the area of an equilateral triangle with side 6 cm. Use a calculator and give your answer correct to three significant figures.

Q3 A plane flies 24 km on a bearing of 020°, then a further 16.7 km on a bearing of 110°. How far is the plane away from its starting point? Use a calculator and given your answer correct to three significant figures.

Answers are on page 208.

Sine, cosine and tangent in right-angled triangles

≫ What do I need to know?

The sides of a right-angled triangle are given special names. These names relate to the angles.

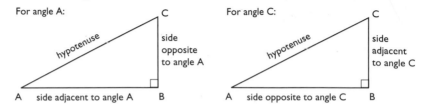

For angle A:

hypotenuse · side opposite to angle A · A · side adjacent to angle A · B · C

For angle C:

hypotenuse · side adjacent to angle C · A · side opposite to angle C · B · C

Sine of an angle

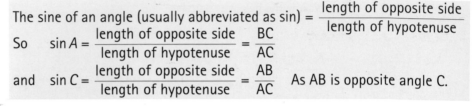

The sine of an angle (usually abbreviated as sin) = $\dfrac{\text{length of opposite side}}{\text{length of hypotenuse}}$

So $\sin A = \dfrac{\text{length of opposite side}}{\text{length of hypotenuse}} = \dfrac{BC}{AC}$

and $\sin C = \dfrac{\text{length of opposite side}}{\text{length of hypotenuse}} = \dfrac{AB}{AC}$ As AB is opposite angle C.

Worked example

Use the sine ratio to find p and q in these right-angled triangles.

a)

p · 10 cm · 42°

b)

8.5 cm · q · 53°

---

a) Using the formula $\sin A = \dfrac{\text{length of opposite side}}{\text{length of hypotenuse}}$:

$\sin 42° = \dfrac{p}{10}$

$p = 10 \times \sin 42°$ Rearranging the formula to make p the subject.

$p = 10 \times 0.669\,130\,6$ As $\sin 42° = 0.669\,130\,6$

$p = 6.691\,306$

$p = 6.69\,\text{cm}$ (3 s.f.)   Including the units and rounding to an appropriate degree of accuracy.

b) Using the formula $\sin A = \dfrac{\text{length of opposite side}}{\text{length of hypotenuse}}$:

$\sin 53° = \dfrac{8.5}{q}$

$q \times \sin 53° = 8.5$ Rearranging the formula to get q on the top line.

$q = \dfrac{8.5}{\sin 53°}$ Rearranging the formula to make q the subject.

$q = \dfrac{8.5}{0.798\,635\,5}$ As $\sin 53° = 0.798\,635\,5$

$q = 10.643\,153$

$q = 10.6\,\text{cm}$ (3 s.f.)   Including the units and rounding to an appropriate degree of accuracy.

Worked example

Use the sine ratio to find the angle θ.

Using the formula $\sin A = \dfrac{\text{length of opposite side}}{\text{length of hypotenuse}}$:

$\sin \theta = \dfrac{2.6}{5.2}$

$\sin \theta = 0.5 \qquad$ As $\dfrac{2.6}{5.2} = 0.5$

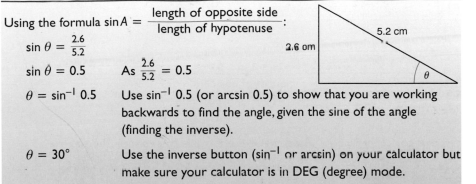

$\theta = \sin^{-1} 0.5 \qquad$ Use $\sin^{-1} 0.5$ (or arcsin 0.5) to show that you are working backwards to find the angle, given the sine of the angle (finding the inverse).

$\theta = 30° \qquad$ Use the inverse button ($\sin^{-1}$ or arcsin) on your calculator but make sure your calculator is in DEG (degree) mode.

≫ Cosine of an angle

The cosine of an angle (usually abbreviated as cos)

$$= \dfrac{\text{length of adjacent side}}{\text{length of hypotenuse}}$$

So $\quad \cos A = \dfrac{\text{length of adjacent side}}{\text{length of hypotenuse}} = \dfrac{AB}{AC}$

and $\quad \cos C = \dfrac{\text{length of adjacent side}}{\text{length of hypotenuse}} = \dfrac{BC}{AC}$

For angle A:

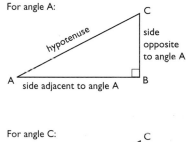

For angle C:

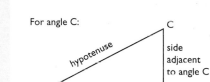

Worked example

Use the cosine ratio to find the angles α and β.

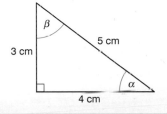

Using the formula $\cos A = \dfrac{\text{length of adjacent side}}{\text{length of hypotenuse}}$:

$\cos \alpha = \dfrac{4}{5}$

$\cos \alpha = 0.8$

$\alpha = \cos^{-1} 0.8 \qquad$ Use $\cos^{-1} 0.8$ (or arccos 0.8) to show that you are working backwards to find the angle, given the cosine of the angle (finding the inverse).

$\alpha = 36.869\,898° \qquad$ Use the inverse button ($\cos^{-1}$ or arccos) on your calculator.

$\alpha = 36.9°$ (3 s.f.)

Similarly $\cos \beta = \dfrac{3}{5}$

$\cos \beta = 0.6$

$\beta = \cos^{-1} 0.6 \qquad$ Use $\cos^{-1} 0.6$ (or arccos 0.6) to show that you are working backwards to find the angle, given the cosine of the angle.

$\beta = 53.130\,102° \qquad$ Use the inverse button ($\cos^{-1}$ or arccos) on your calculator.

$\beta = 53.1°$ (3 s.f.)

Worked example

Use the cosine ratio to find the lengths indicated on the right-angled triangles.

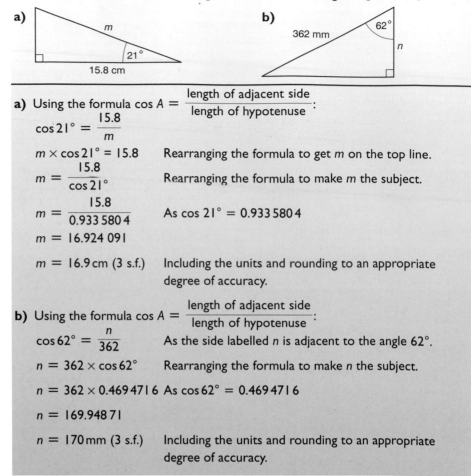

a) Using the formula $\cos A = \dfrac{\text{length of adjacent side}}{\text{length of hypotenuse}}$:

$\cos 21° = \dfrac{15.8}{m}$

$m \times \cos 21° = 15.8$ Rearranging the formula to get m on the top line.

$m = \dfrac{15.8}{\cos 21°}$ Rearranging the formula to make m the subject.

$m = \dfrac{15.8}{0.933\,580\,4}$ As $\cos 21° = 0.933\,580\,4$

$m = 16.924\,091$

$m = 16.9\,\text{cm}$ (3 s.f.)   Including the units and rounding to an appropriate degree of accuracy.

b) Using the formula $\cos A = \dfrac{\text{length of adjacent side}}{\text{length of hypotenuse}}$:

$\cos 62° = \dfrac{n}{362}$ As the side labelled n is adjacent to the angle 62°.

$n = 362 \times \cos 62°$ Rearranging the formula to make n the subject.

$n = 362 \times 0.469\,471\,6$ As $\cos 62° = 0.469\,471\,6$

$n = 169.9487\,1$

$n = 170\,\text{mm}$ (3 s.f.)   Including the units and rounding to an appropriate degree of accuracy.

HINT

▶ The value of n might also be found by using $\sin 28° = \dfrac{n}{362}$ where 28° is the missing angle.

≫ Tangent of an angle

The tangent of an angle (usually abbreviated as tan) = $\dfrac{\text{length of opposite side}}{\text{length of adjacent side}}$

So $\tan A = \dfrac{\text{length of opposite side}}{\text{length of adjacent side}} = \dfrac{BC}{AB}$

and $\tan C = \dfrac{\text{length of opposite side}}{\text{length of adjacent side}} = \dfrac{AB}{BC}$

HINT

▶ Once the angle α had been found then it is possible to find β by using the fact that $\alpha + \beta + 90° = 180°$ (the angle sum of a triangle = 180°).

For angle A:

hypotenuse
C
side opposite to angle A
A   side adjacent to angle A   B

For angle C:

hypotenuse
C
side adjacent to angle C
A   side opposite to angle C   B

Worked example

Given that $AD = 30\,cm$, $BC = 10\,cm$ and $\angle DBC = 61°$ find:

a) CD

b) $\angle DAC$.

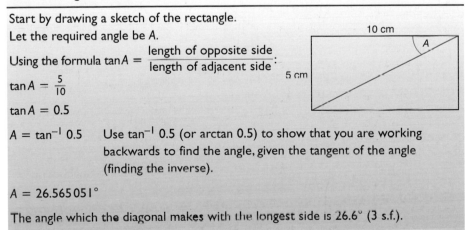

a) For the right-angled triangle DBC, using the formula

$$\tan DBC = \frac{\text{length of opposite side}}{\text{length of adjacent side}}:$$

$\tan 61° = \dfrac{CD}{10}$

$CD = 10 \times \tan 61°$ Rearranging the formula to make CD the subject.

$CD = 10 \times 1.804\,047\,8$ As $\tan 61° = 1.804\,047\,8$

$CD = 18.040\,478$

$CD = 18.0\,cm$ (3 s.f.)   Including the units and rounding to an appropriate degree of accuracy.

b) For the right-angled triangle DAC, using the formula

$$\sin A = \frac{\text{length of opposite side}}{\text{length of hypotenuse}}:$$

$\sin DAC = \dfrac{18.040\,478}{30}$ Using the original value of CD and not the rounded value.

$\sin DAC = 0.601\,349\,3$

$\angle DAC = \sin^{-1} 0.601\,349\,3$

$\angle DAC = 36.966\,593°$ Use the inverse button ($\sin^{-1}$ or arcsin) on your calculator.

$\angle DAC = 37.0°$ (3 s.f.)   Rounding to an appropriate degree of accuracy.

> **＋ HINT**
>
> ▶ Learn how to use the memory buttons on your calculator. You can use more exact values, and it makes you less likely to make a keying error.

Worked example

A rectangle measures 10 cm by 5 cm. What angle does the diagonal make with the longer sides?

Start by drawing a sketch of the rectangle.

Let the required angle be A.

Using the formula $\tan A = \dfrac{\text{length of opposite side}}{\text{length of adjacent side}}:$

$\tan A = \dfrac{5}{10}$

$\tan A = 0.5$

$A = \tan^{-1} 0.5$ Use $\tan^{-1} 0.5$ (or arctan 0.5) to show that you are working backwards to find the angle, given the tangent of the angle (finding the inverse).

$A = 26.565\,051°$

The angle which the diagonal makes with the longest side is 26.6° (3 s.f.).

≫ Angles of elevation and depression

The **angle of elevation** is the angle up from the horizontal.

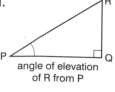

angle of elevation
of R from P

The **angle of depression** is the angle down from the horizontal.

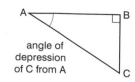

angle of
depression
of C from A

Worked example

An object is situated 25 metres from the foot of a mast of height 32 metres. What is the angle of depression of the object from the top of the mast?

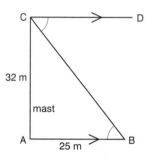

Start by drawing a diagram of the situation.

The angle of depression of the object from the top of the mast is $\angle$DCB.

$\angle$DCB = $\angle$ABC   As these are alternate angles between parallel lines AB and CD.

$\tan ABC = \dfrac{32}{25}$ As the tangent of an angle = $\dfrac{\text{length of opposite side}}{\text{length of adjacent side}}$.

$\tan ABC = 1.28$

$\angle ABC = \tan^{-1} 1.28$ Use the inverse button ($\tan^{-1}$ or arctan) on your calculator.

$\angle ABC = 52.001\,268°$

The angle of depression is 52.0° (3 s.f.).

CHECK YOURSELF QUESTIONS

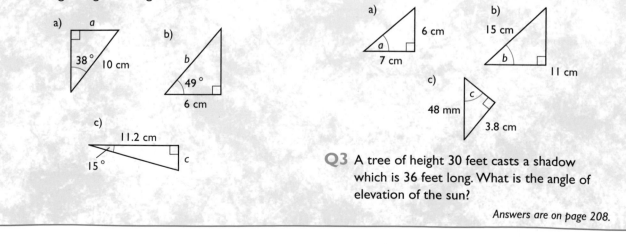

Q1 Find the lengths of the sides in the following right-angled triangles.

a)

38°   10 cm   *a*

b)

b   49°   6 cm

c)

11.2 cm   15°   *c*

Q2 Find the angles in the following triangles.

a)

a   6 cm   7 cm

b)

15 cm   *b*   11 cm

c)

48 mm   *c*   3.8 cm

Q3 A tree of height 30 feet casts a shadow which is 36 feet long. What is the angle of elevation of the sun?

Answers are on page 208.

Lengths, areas and volumes

≫ What do I need to know?

You will need to know about the areas of triangles and quadrilaterals, as given in Revision session 1, *Geometric terms*.

The formulae for the area of a trapezium and the volume of a prism will be given on the examination paper at the Intermediate level.

- **Area of trapezium** $= \frac{1}{2}(a + b)h$

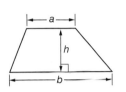

- **Volume of prism**
  $=$ area of cross-section $\times$ length

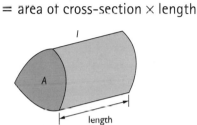

≫ Other useful formulae

- **Circumference of circle**
  $= \pi \times$ diameter

  $= 2 \times \pi \times$ radius

- **Area of circle**
  $= \pi \times$ (radius)^2

- **Area of parallelogram**
  $=$ base $\times$ height

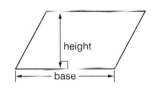

- **Volume of cuboid**
  $=$ length $\times$ width $\times$ height

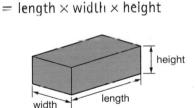

- **Volume of cylinder** $= \pi r^2 h$

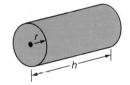

Worked example

Find the perimeter and area of a circle of diameter 10 metres.

Perimeter of a circle $= \pi \times d$	Diameter of the circle $= d$
$\qquad = \pi \times 10$	
$\qquad = 10\pi$ metres	Leaving the answer in terms of π.
Area of a circle $\qquad = \pi \times r^2$	Area of a circle $= \pi r^2$
$\qquad = \pi \times 5^2$	Radius $= \frac{1}{2} \times$ diameter
$\qquad = 25\pi$ square metres	

+ HINT

▶ On the 'calculator-allowed' paper, you can use the value of π to get the answers of 31.4 m (to 3 s.f.). and 78.5 m^2 (to 3 s.f.) respectively.

Worked example

Find the volume of the prism with dimensions as shown.

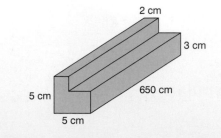

Volume of a prism = area of cross-section × length.

Area of cross-section $= 5 \times 3 + 2 \times 2$

$\qquad\qquad\qquad\quad = 15 + 4$

$\qquad\qquad\qquad\quad = 19 \, cm^2$

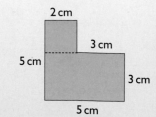

Volume of a prism = area of cross-section × length

$\qquad\qquad = 19 \times 65$ $\qquad$ Writing all lengths in the same units with 10 mm = 1 cm.

$\qquad\qquad = 1235 \, cm^3$

$\qquad\qquad = 1240 \, cm^3$ (3 s.f.)

≫ Units for length, area and volume

Length, area and volume can all be identified by looking at the formulae or units, as area is two-dimensional (the product of two lengths) and volume is three-dimensional (the product of three lengths).

By ignoring constants (including π) you should be able to identify length, area and volume as shown in the table.

Lengths	Areas	Volumes
d	lb	lbh
$2\pi r$	$\frac{1}{2}lb$	$\frac{4}{3}\pi r^3$
	$\frac{1}{2}(a+b)h$	
	πr^2	$\pi r^2 h$
	$2\pi rh$	
	πrl	$\frac{1}{3}\pi r^2 h$
	πab	

✚ HINT

▶ The sum of two lengths is still a length and the sum of two areas is still an area etc.

CHECK YOURSELF QUESTIONS

Q1 Calculate the areas of the shaded parts of the following shapes.

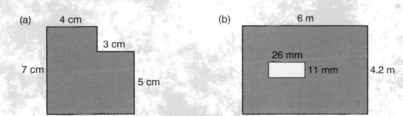

(a) 4 cm, 3 cm, 7 cm, 5 cm

(b) 6 m, 26 mm, 11 mm, 4.2 m

Q2 A circular disc is cut from a square of side 7 cm to leave a minimum amount of waste. What is the area of the waste and what is this as a percentage of the original area?

Q3 A circular pond has a surface area of 400 m². Calculate the diameter of the pond to an appropriate degree of accuracy.

Q4 A washer has an outside diameter of 12 mm and an inside diameter of 6 mm. Calculate the area of cross-section of the washer, leaving your answer in terms of π.

Q5 The diameter of a £1 coin is 22 mm. The coin is 3 mm thick. Work out the volume of the coin, giving your answer in terms of π.

Q6 Identify the following as length, area or volume.

a) $2\pi r(r+h)$ b) $\frac{\theta}{360} \times 2\pi r$

c) $\frac{\pi}{4}d^2 h$ d) $\sqrt{r^2+h^2}$

Answers are on page 209.

Angle properties of circles

≫ Parts of a circle

This diagram shows the main parts of a circle and the names given to them.

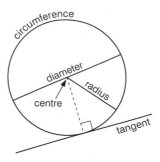

≫ Angles in circles

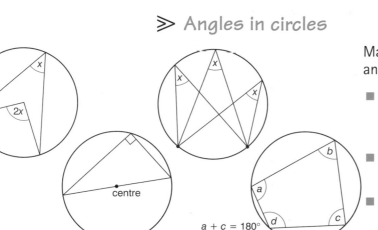

$a + c = 180°$
$b + d = 180°$

Make sure that you know the following angle properties.

- The angle subtended by an arc (or chord) at the centre is twice that subtended at the circumference.

- Angles subtended by the same arc (or chord) are equal.

- The angle in a semicircle is always 90°.

- The opposite angles of a cyclic quadrilateral are **supplementary** (they add up to 180°).

Worked example

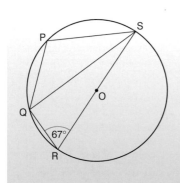

PQRS is a quadrilateral inscribed in a circle centre O. Find, giving reasons for your answers, ∠QSR and ∠QPS.

∠SQR = 90°	As RS is a diameter and the angle in a semicircle is 90°.
∠QSR = 180° − (67° + 90°) = 23°	As the angles of a triangle add up to 180°.
∠QPS = 180° − 67°	As the opposite angles of a cyclic quadrilateral PQRS add up to 180°.
= 113°	

Worked example

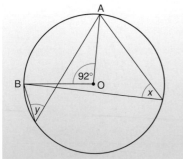

The angle subtended by the arc AB at the centre of the circle is 92°. Find the values of x and y.

x = 46°	As the angle subtended by the arc AB at the centre is twice that subtended at the circumference.
y = 46°	As the angle subtended by the arc AB at the centre is twice that subtended at the circumference.
or else ...	As the angles subtended by the arc AB at the circumference are equal and x = 46°, so y = 46°.

≫ Chord properties

A **chord** is a straight line joining two points on the circumference of a circle.

You need to know the following chord properties.

■ A perpendicular from the centre of a circle to a chord bisects the chord. Conversely, a perpendicular bisector of a chord passes through the centre of the circle.

■ Chords that are equal in length are equidistant from the centre of the circle. Conversely, chords that are equidistant from the centre of a circle are equal in length.

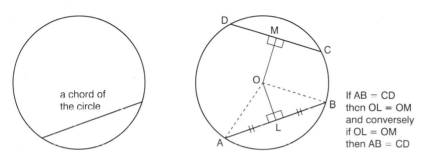

a chord of the circle

If AB = CD
then OL = OM
and conversely
if OL = OM
then AB = CD

≫ Tangent properties

A **tangent** is a straight line which touches a circle at one point only.

You need to know the following tangent properties.

■ A tangent to a circle is perpendicular to the radius at the point of contact.

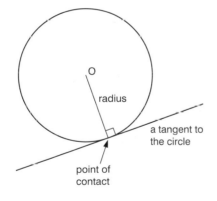

radius

a tangent to the circle

point of contact

■ Tangents to a circle from an external point are equal in length.

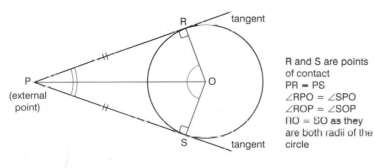

P
(external point)

tangent

tangent

R and S are points
of contact
PR = PS
∠RPO = ∠SPO
∠ROP = ∠SOP
RO = SO as they
are both radii of the
circle

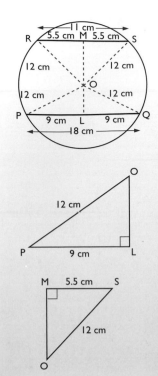

Worked example

Two chords PQ and RS are parallel to each other on opposite sides of a circle of radius 12 cm. If PQ = 18 cm and RS = 11 cm, find the distance between the chords.

Show this information on a diagram and use the chord properties to show the respective lengths.

For the right-angled triangle OPL:

$12^2 = 9^2 + OL^2$ Applying Pythagoras' theorem to the right-angled triangle.

$144 = 81 + OL^2$

$OL^2 = 63$

$OL = 7.937\,254$

For the right-angled triangle OMS:

$12^2 = 5.5^2 + OM^2$ Applying Pythagoras' theorem to the right-angled triangle.

$144 = 30.25 + OM^2$

$OM^2 = 113.75$

$OM = 10.665\,365$

Distance between the two chords $= OL + OM$

$= 7.937\,254 + 10.665\,365$

$= 18.602\,619$

$= 18.6$ cm (3 s.f.)

? CHECK YOURSELF QUESTIONS

Q1 For the following circles, where O marks the centre of the circle, find the missing angles.

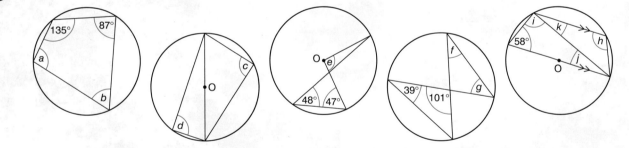

Q2 In the triangle ABC, AB is a diameter of length 11 cm and AC measures 3.5 cm.

Find **a)** BC   **b)** ∠BAC   **c)** ∠ABC.

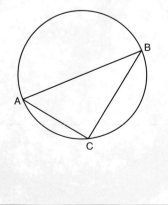

Q3 A chord AB is drawn on a circle of radius 6 cm. If the chord is 4.5 cm from the centre of the circle, calculate the length of the chord.

Q4 Given that PR and PS are tangents to a circle and the points of contact are R and S respectively, find:

a) ∠POS   **b)** ∠OPR   **c)** ∠OPS.

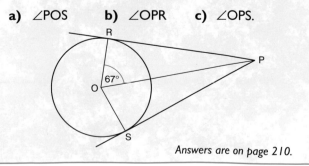

Answers are on page 210.

Further areas and volumes

≫ What do I need to know?

Make sure that you have worked through areas of triangles and quadrilaterals as given in Revision sessions 1, *Geometric terms* and 12, *Lengths, areas and volumes*.

You will be given the formulae for the volume of a prism, the volume of a sphere, the volume of a cone, the surface area of a sphere and the curved surface of a cone on the Higher examination paper.

- Volume of cylinder = $\pi r^2 h$
- Curved surface of cylinder = $2\pi rh$

- Volume of sphere = $\frac{4}{3}\pi r^3$
- Surface area of sphere = $4\pi r^2$

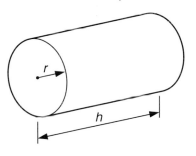

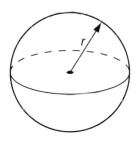

- Volume of cone = $\frac{1}{3}\pi r^2 h$
- Curved surface area of cone = πrl

- Volume of a prism
  = area of cross-section × length

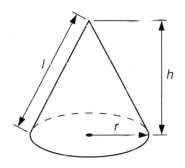

A is the area of cross-section

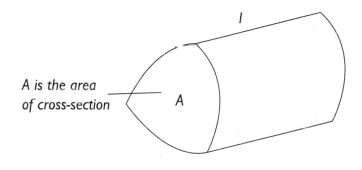

≫ Area and volume

Worked example

The dimensions of a cone are given in the diagram.

Calculate:

a) the area of the curved surface of the cone

b) the volume of the cone

giving your answers in terms of π.

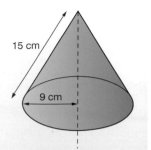

a) Curved surface area of cone
$= \pi r l$
$= \pi \times 9 \times 15$ As radius $= 9$ cm and slant height $= 15$ cm.
$= \pi \times 135$
$= 135\pi$ cm^2

b) Volume of cone $= \frac{1}{3}\pi r^2 h$ Where h is the perpendicular height.

Using Pythagoras' theorem:
$l^2 = r^2 + h^2$
$15^2 = 9^2 + h^2$
$225 = 81 + h^2$
$h^2 = 225 - 81 = 144$
$h = 12$ Taking square roots on both sides.

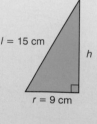

Volume of cone $= \frac{1}{3}\pi r^2 h$

$= \frac{1}{3} \times \pi \times 9^2 \times 12$
$= \pi \times 324$
$= 324\pi$ cm^3

≫ Scale factors of length, area and volume

Two solids are **similar** if the ratios of their corresponding linear dimensions are equal.

In general:
- the corresponding **areas of similar solids are proportional to the squares of their linear dimensions**

- the corresponding **volumes of similar solids are proportional to the cubes of their linear dimensions.**

So if an object is enlarged by a scale factor of s:

- lengths are multiplied by s

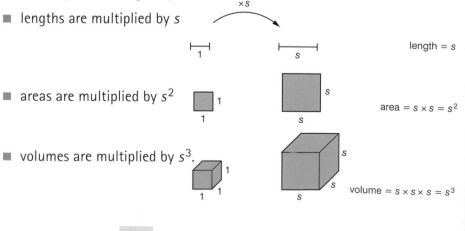

- areas are multiplied by s^2

- volumes are multiplied by s^3.

Worked example

The ratio of the surface areas of two similar cylinders is $16 : 25$. Calculate the ratio of their volumes and work out the volume of the larger cylinder given that the smaller cylinder has a volume of $540 \, \text{mm}^2$.

If the ratio of the lengths is $\quad x : y$

then the ratio of the areas is $\quad x^2 : y^2$

and the ratio of the volumes is $\quad x^3 : y^3$

Here $\quad x^2 : y^2 \quad = \quad 16 : 25$

so $\quad x : y \quad = \quad 4 : 5$

and $\quad x^3 : y^3 \quad = \quad 4^3 : 5^3$

$\quad\quad\quad\quad\quad\quad = \quad 64 : 125$

So the ratio of the volumes is $64 : 125$.

$64 : 125 = 1 : \frac{125}{64}$ Rewriting as an equivalent ratio.

$\quad = 1 \times 540 : \frac{125}{64} \times 540$ Rewriting as an equivalent ratio with 540 on the left-hand side.

$\quad = 540 : 1054.6875$

The volume of larger cylinder is $1050 \, \text{mm}^3$ (3 s.f.).

? CHECK YOURSELF QUESTIONS

Q1 The surface area of a sphere is $36\pi \, \text{cm}^2$. Calculate the volume of the sphere, giving your answer in terms of π.

Q2 The corresponding lengths of two similar solids are in the ratio $3 : 5$. What is the ratio of:

a) their surface areas

b) their volumes?

Q3 The curved surface area of a cone is $1165 \, \text{mm}^2$. What is the curved surface area of a similar cone with height three times the height of the original cone?

Answers are on page 211.

Three-dimensional trigonometry

≫ What do I need to know?

For the Higher paper examination you will be required to work with trigonometry in three dimensions. For this work, it is useful to identify right-angled triangles and show them diagrammatically to answer the questions.

≫ Working in three dimensions

Worked example

The pyramid OABCD has a square base of length 15 cm and a vertical height of 26 cm. Calculate:

a) the length OA

b) the angle OAC.

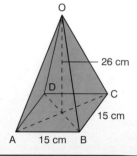

---

a) Consider the base ABCD (where X is the centre of the square base).

$AC^2 = 15^2 + 15^2$ Using Pythagoras' theorem on the right-angled triangle ABC.

$AC^2 = 225 + 225$

$AC^2 = 450$

$AC = 21.213\,203$ Taking square roots on both sides.

$AX = \frac{1}{2} \times AC$

$AX = 10.606\,602$

Now consider triangle OAX.

$OA^2 = AX^2 + OX^2$ Using Pythagoras' theorem on the right-angled triangle AXO where OX is the vertical height.

$OA^2 = 10.606\,602^2 + 26^2$

$OA^2 = 788.5$

$OA = 28.080\,242$ Taking square roots on both sides.

$OA = 28.1$ cm (3 s.f.) Rounding to an appropriate degree of accuracy.

b) The angle OAC is the same as the angle OAX in the previous diagram.

$\tan OAC = \frac{OX}{AX}$ As $\tan \theta = \dfrac{\text{length of opposite side}}{\text{length of adjacent side}}$

$\tan OAC = \frac{26}{10.606\,602}$

$\tan OAC = 2.451\,303\,4$

$\angle OAC = 67.807\,182°$ Using $\tan^{-1}$ or arctan to find the angle.

$\angle OAC = 67.8°$ (3 s.f.) Rounding to an appropriate degree of accuracy.

Q1 Find the following lengths in the cuboid shown in the diagram. Give your answers in surd form, expressing them in their lowest terms where possible.

a) AF   **b)** AC   **c)** AG

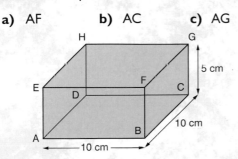

Q2 Find the area of the plane ADGF in this diagram. Give your answer in surd form, expressing it in its lowest terms.

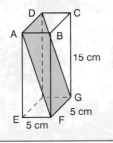

Q3 Find the angle which the line EC in the diagram makes with the base of the cube.

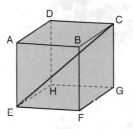

Q4 An electric pylon XY is 30 m high. From a point S, due south of the pylon, the angle of elevation of the top of the pylon is 26° and from a point W, due west of the pylon, the angle of elevation is 32°. Find the distance WS.

Answers are on page 211.

Sine, cosine and tangent for angles of any size

≫ Finding sine, cosine, tangent

The sine, cosine and tangent can be found for any angle by using a calculator or else using the properties of sine, cosine and tangent curves, as shown in the following graphs.

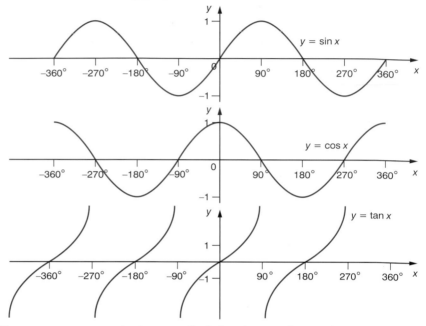

You can use your calculator to find the sine, cosine and tangent of any angle, although the reverse process is not so easy as the calculator only gives answers in a specified range.

Worked example

Find the values of θ in the range $^-360° \leqslant \theta \leqslant 360°$ such that $\tan\theta = 1$.

Using the calculator:

$\tan\theta = 1$

$\theta = \tan^{-1} 1$

$\theta = 45°$

However, this is only one of many solutions, as can be seen from the following graph of $\tan\theta$.

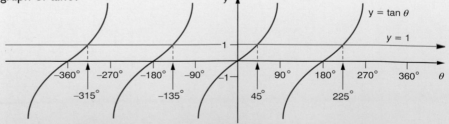

From the graph you can see that the solutions in the range $^-360° \leqslant \theta \leqslant 360°$ are $^-315°$, $^-135°$, $45°$ and $225°$.

You can check these on your calculator by evaluating $\tan^-315°$, $\tan^-135°$ and $\tan 225°$.

CHECK YOURSELF QUESTIONS

Q1 Find all the values of *x* between 0° and 360° for which:

a) sin *x* = 0.6

b) sin *x* = ⁻0.6

c) cos *x* = 0.2

d) tan *x* = ⁻2.5.

Q2 Solve these equations for ⁻360° ≤ *x* ≤ 360°.

a) tan *x* = 2

b) sin *x* = 0.525

Q3 Solve these equations for 0° ≤ *x* ≤ 360°.

a) sin *x* ≥ 0.5

b) cos *x* < 0.5

Answers are on page 212.

Sine and cosine rules

≫ Why use sine and cosine rules?

You can use the sine, cosine and tangent ratios, introduced in Revision session 16, *Sine, cosine and tangent in right-angled triangles* to **solve** right-angled triangles, which means finding all the missing angles and sides. You need the sine and cosine rules to solve triangles that are not right-angled.

Use the **sine rule** when you are working with **two sides and two angles**.

Use the **cosine rule** when you are working with **three sides and one angle**.

≫ Sine rule

The sine rule states that: $\dfrac{a}{\sin A} = \dfrac{b}{\sin B} = \dfrac{c}{\sin C}$

Sometimes it is useful to use the alternative form: $\dfrac{\sin A}{a} = \dfrac{\sin B}{b} = \dfrac{\sin C}{c}$

≫ Cosine rule

The cosine rule states that:

$a^2 = b^2 + c^2 - 2bc\cos A$

or $\cos A = \dfrac{b^2 + c^2 - a^2}{2bc}$

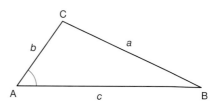

Area of a triangle

The sine and cosine rules can be extended to find the area of a triangle so that:

area of a triangle = $\frac{1}{2}ab\sin C$

where C is the 'included' angle between sides a and b.

Worked example

Find the angle A in the triangle ABC where AB is 11 cm, BC is 17.5 cm and angle C is 36°.

The information involves two sides and two angles, so use the sine rule.

$\dfrac{a}{\sin A} = \dfrac{b}{\sin B}$

$\dfrac{\sin A}{a} = \dfrac{\sin B}{b}$ Reciprocating both sides.

$\dfrac{\sin A}{17.5} = \dfrac{\sin 36°}{11}$

$\sin A = 17.5 \times \dfrac{\sin 36°}{11}$ Multiplying both sides by 17.5.

$\sin A = 0.9351129$

$A = 69.246393°$ Using the inverse button (sin^{-1} or arcsin) on your calculator.

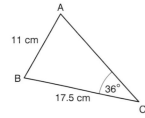
11 cm

17.5 cm

36°

Unfortunately the value of angle A = 69.246 392° is not unique since there is another possible value that satisfies sin A = 0.935 112 9.

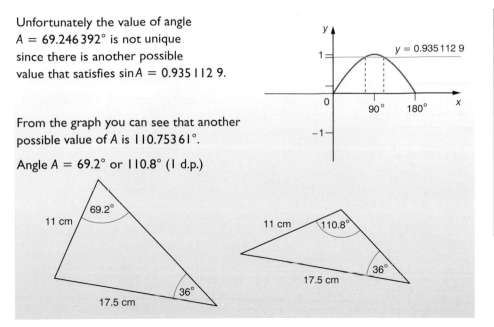

y = 0.935 112 9

From the graph you can see that another possible value of A is 110.753 61°.

Angle A = 69.2° or 110.8° (1 d.p.)

HINT

▸ Whenever you use the sine rule you need to be aware of the possibility that two solutions exist. This problem does not arise when using the cosine rule and can be avoided when using the sine rule by finding the smallest angle of a triangle first (if possible).

Worked example

Find the angle P and the area of the triangle.

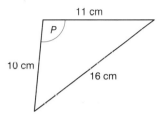

Since the information involves three sides and one angle, use the cosine rule.

$$\cos A = \frac{b^2 + c^2 - a^2}{2bc}$$

$$\cos P = \frac{10^2 + 11^2 - 16^2}{2 \times 10 \times 11}$$ Substituting the given lengths.

$$\cos P = \frac{100 + 121 - 256}{220}$$

$$\cos P = \frac{^-35}{220}$$

$\cos P = {}^-0.159\,090\,9$ The negative value shows that the angle P is obtuse.

$P = 99.154\,133°$ Using the inverse button (cos^{-1} or arccos) on the calculator.

$P = 99.2°$ (1 d.p.)

To find the area of the triangle use:

area of a triangle $= \frac{1}{2}ab\sin C$

$= \frac{1}{2} \times 10 \times 11 \times \sin 99.154\,133°$ Remembering to use the most accurate value of P.

$= 54.299\,517$

$= 54.3\,cm^2$ (3 s.f.)   Rounding to an appropriate degree of accuracy.

Q1 Calculate the angles and sides marked with letters in the following triangles.

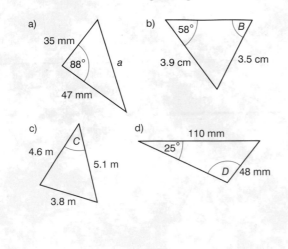

a)

35 mm

88°

a

47 mm

b)

58°

B

3.9 cm

3.5 cm

c)

C

4.6 m

5.1 m

3.8 m

d)

110 mm

25°

D

48 mm

Q2 Calculate the areas of the given shapes.

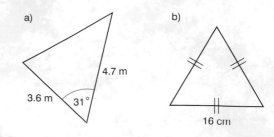

a)

4.7 m

3.6 m

31°

b)

16 cm

Q3 A ship leaves port and sails on a bearing of 035°. After 11 km the ship changes direction and sets sail on a bearing of 067°. The ship sails a distance of 16.5 km on this bearing. What is the distance and bearing of the ship from the port?

Answers are on page 214.

Arc, sector and segment

≫ Parts of a circle

Make sure you know the definitions of parts of the circle, given in the diagrams below. You will find the formulae and reminders in the rest of this session useful, too.

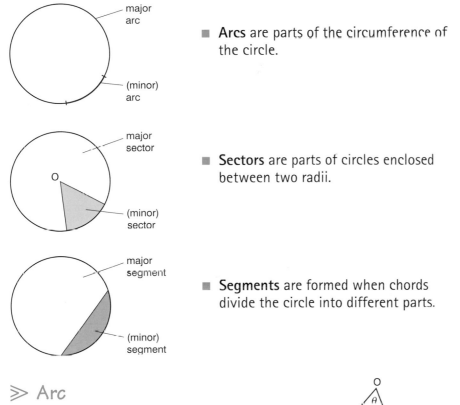

■ **Arcs** are parts of the circumference of the circle.

■ **Sectors** are parts of circles enclosed between two radii.

■ **Segments** are formed when chords divide the circle into different parts.

≫ Arc

An arc is a part of the circumference of a circle.

Arc length $= \dfrac{\text{angle subtended at centre}}{360} \times 2\pi r$

Arc length $= \dfrac{\theta}{360} \times 2\pi r$

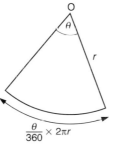

$\dfrac{\theta}{360} \times 2\pi r$

Worked example

Find the perimeter of this shape.

Perimeter $=$ arc $+$ radius $+$ radius

$= \dfrac{320}{360} \times 2\pi r + r + r$

$= \dfrac{320}{360} \times 2 \times \pi \times 4 + 4 + 4$

$= 30.340\,214$

$= 30.3\,\text{cm}$ (3 s.f.)

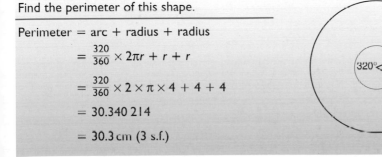

≫ Sector

A sector is the area enclosed between an arc and two radii.

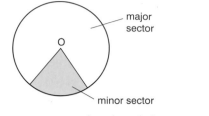

$$\text{Sector area} = \frac{\text{angle subtended at centre}}{360} \times \pi r^2$$

$$\text{Sector area} = \frac{\theta}{360} \times \pi r^2$$

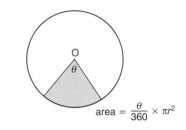

$$\text{area} = \frac{\theta}{360} \times \pi r^2$$

≫ Segment

A segment is the area enclosed between an arc and a chord.

The following example illustrates how to find the area of a segment.

Worked example

Find the area of the segment shaded in this circle of radius 5 cm.

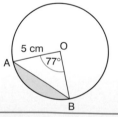

Area of segment = area of sector AOB − area of triangle AOB

$$\text{Area of sector AOB} = \frac{77}{360} \times \pi r^2$$

$$= \frac{77}{360} \times \pi \times 5^2$$

$$= 16.798\,794 \text{ cm}^2$$

$$\text{Area of triangle AOB} = \frac{1}{2} ab\sin\theta$$

$$= \frac{1}{2} \times 5 \times 5 \times \sin 77° \qquad \text{Where } a \text{ and } b \text{ are equal to the radius of the circle.}$$

$$= 12.179\,626 \text{ cm}^2$$

Area of segment = area of sector AOB − area of triangle AOB

$$= 16.798\,794 - 12.179\,626$$

$$= 4.619\,168$$

$$= 4.62 \text{ cm}^2 \text{ (3 s.f.)}$$

Q1 Find the arc length and sector area for the circle below, giving your answers in terms of π.

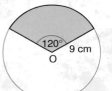

120° 9 cm
O

Q2 Calculate the segment area in this diagram.

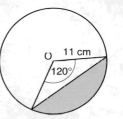

O 11 cm
120°

Q3 The arc length between two points A and B on the circumference of a circle is 8.6 cm. What is the angle subtended at the centre of the circle if the radius of the circle is 12 cm?

Q4 The sector of a circle is folded to make a cone of slant height 16 cm and base radius 12 cm. Calculate the arc length and the angle of the original sector.

Answers are on page 215.

Further angle properties of circles

≫ What do I need to know?

In Revision session 13, *Angle properties of circles* you used the angle properties. At the Higher level, you need to be able to provide proofs for them.

≫ Angle properties

Worked example

Prove that the opposite angles in a cyclic quadrilateral add up to 180°.

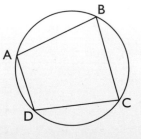

Draw the radii OA and OC.

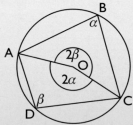

Let $\angle ABC = \alpha$

Let $\angle ADC = \beta$

Then the lower $\angle AOC = 2\alpha$ As the angle subtended by the arc AC at the centre is twice the angle subtended at the circumference.

and the upper $\angle AOC = 2\beta$ As the angle subtended by the arc AC at the centre is twice the angle subtended at the circumference.

But $2\alpha + 2\beta = 360°$ As angles at a point add up to 360°.

so $2(\alpha + \beta) = 360°$ Factorising the left-hand side.

and $\alpha + \beta = 180°$ Dividing both sides by 2.

As α and β are opposite angles of a cyclic quadrilateral, the opposite angles in a cyclic quadrilateral add up to 180°.

≫ The alternate segment theorem

■ The alternate segment theorem states that the angle between a tangent and a chord equals the angle subtended by the chord in the alternate segment.

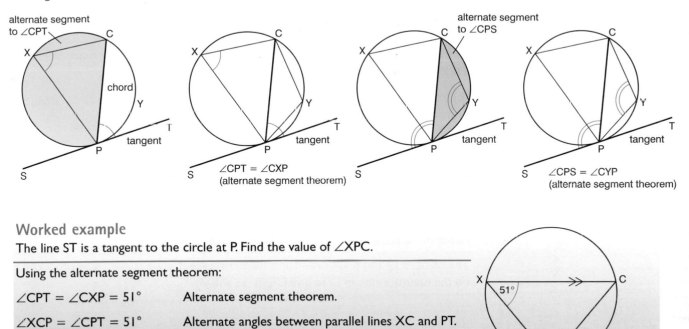

∠CPT = ∠CXP
(alternate segment theorem)

∠CPS = ∠CYP
(alternate segment theorem)

Worked example

The line ST is a tangent to the circle at P. Find the value of ∠XPC.

Using the alternate segment theorem:

∠CPT = ∠CXP = 51°   Alternate segment theorem.

∠XCP = ∠CPT = 51°   Alternate angles between parallel lines XC and PT.

∠XPC = 180° − (51° + 51°) Angles of triangle XCP add up to 180°.

∠XCP = 78°

CHECK YOURSELF QUESTIONS

Q1 Prove the alternate segment theorem.

Q2 Given that angle CPT = 61° find:

    a) ∠PXC   **b)** ∠PYC.

Q3 Given that angle CDP = 38° find:

    a) ∠COP   **b)** ∠CPT   **c)** ∠OCP.

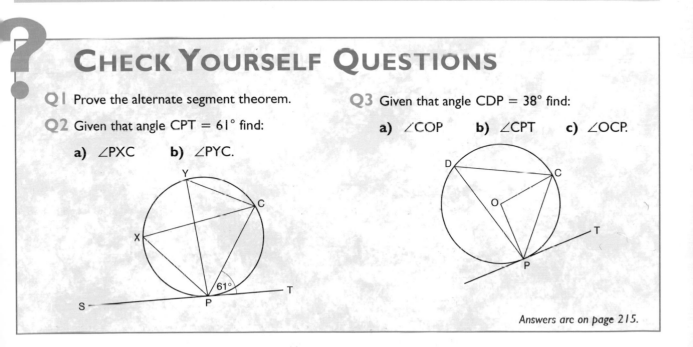

Answers are on page 215.

Enlargement with negative scale factors

≫ What does a negative scale factor mean?

An enlargement with a negative scale factor means that the enlargement is situated on the opposite side of the centre of enlargement.

Enlargements with positive scale factors were discussed in Revision session 7, *Transformations*.

≫ Working with negative scale factors

Worked example

The triangle PQR with vertices P(2, 2), Q(4, 2) and R(2, 6) is enlarged with a scale factor of ⁻2 about the origin. Draw PQR and hence P′Q′R′.

The points P, Q and R are drawn and the enlargement, scale factor ⁻2, is produced on the opposite side of O to give P′Q′R′ as shown.

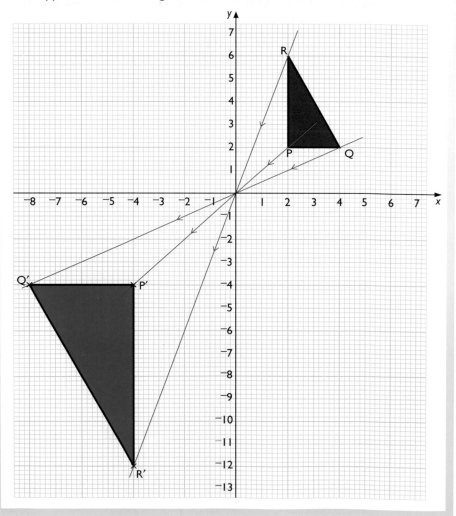

CHECK YOURSELF QUESTIONS

Q1 Draw the image of the triangle PQR after an enlargement, scale factor ⁻2, about the given centre of enlargement O.

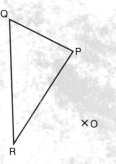

Q2 On graph paper, draw the triangle R with vertices (1, 1), (3, 1) and (1, 2).

Enlarge the triangle R with scale factor 3 and centre (0, 0) to form S.

Enlarge the triangle S with scale factor ⁻$\frac{1}{3}$ and centre (0, 0) to form T.

What single transformation maps R onto T?

Q3 Find the centre of enlargement and the scale factor of this enlargement.

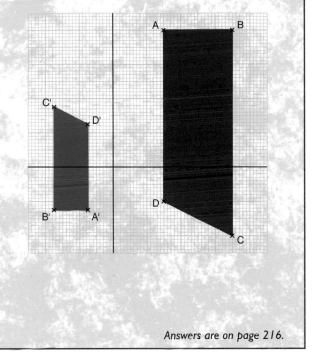

Answers are on page 216.

REVISION SESSION 21
HIGHER

Vectors and vector properties

≫ What is a vector?

A **vector** is a quantity which has magnitude (length) and direction (as indicated by the arrow).

Displacement, velocity, acceleration, force and **momentum** are all examples of vectors.

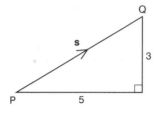

≫ Representing a vector

The vector on the left can be represented in a number of ways:

- **PQ** or you can write $\overrightarrow{PQ}$
- **s** or you can write s
- as a column vector $\begin{pmatrix} 5 \\ 3 \end{pmatrix}$ as in Revision session 7, *Transformations*.

≫ Equal vectors

Two vectors are equal if they have the same magnitude and direction, which means that they are the same length and they are parallel.

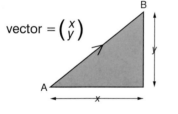

vector $= \begin{pmatrix} x \\ y \end{pmatrix}$

≫ Components of a vector

The components of a vector are usually described in terms of:

- the number of units moved in the *x*-direction
- the number of units moved in the *y*-direction.

These units are best expressed as a column vector, $\begin{pmatrix} \text{change in } x\text{-value} \\ \text{change in } y\text{-value} \end{pmatrix}$.

≫ Adding and subtracting vectors

Vectors can be added or subtracted by placing them end to end, so that the arrows point in the same direction or lead on from one to the next.

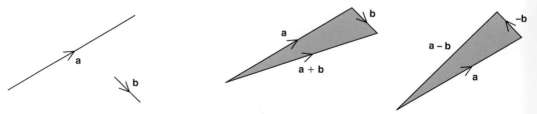

a − b is the same as **a + (−b)** where **−b** is the same as **b** but acts in the opposite direction.

An easier way to add vectors is to write them as column vectors so that if

$a = \begin{pmatrix} 5 \\ 3 \end{pmatrix}$ and $b = \begin{pmatrix} 1 \\ -1 \end{pmatrix}$ then:

$a + b = \begin{pmatrix} 5 \\ 3 \end{pmatrix} + \begin{pmatrix} 1 \\ -1 \end{pmatrix} = \begin{pmatrix} 6 \\ 2 \end{pmatrix}$

$a - b = \begin{pmatrix} 5 \\ 3 \end{pmatrix} - \begin{pmatrix} 1 \\ -1 \end{pmatrix} = \begin{pmatrix} 4 \\ 4 \end{pmatrix}$

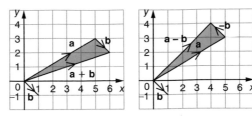

≫ Magnitude of a vector

You can use Pythagoras' theorem to find the magnitude (length) of a vector.

Suppose that $\overrightarrow{AB} = \begin{pmatrix} x \\ y \end{pmatrix}$.

Then the length of the vector $\overrightarrow{AB} = \sqrt{x^2 + y^2}$ and you can write

$|\overrightarrow{AB}| = \sqrt{x^2 + y^2}$

where the two vertical lines stand for 'magnitude of' or 'length of'.

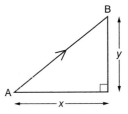

Worked example

Given that $p = \begin{pmatrix} 3 \\ 4 \end{pmatrix}$ and $q = \begin{pmatrix} 2 \\ -1 \end{pmatrix}$, find $p + q$ and $|p + q|$.

$p + q = \begin{pmatrix} 3 \\ 4 \end{pmatrix} + \begin{pmatrix} 2 \\ -1 \end{pmatrix}$

$= \begin{pmatrix} 5 \\ 3 \end{pmatrix}$

$|p + q| = \sqrt{5^2 + 3^2}$

$= \sqrt{34}$ units

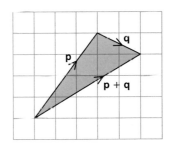

≫ Multiplication of a vector

Vectors cannot be multiplied by other vectors but they can be multiplied by a constant (sometimes called **scalar multiplication**).

Worked example

Given that $p = \begin{pmatrix} 3 \\ 4 \end{pmatrix}$ and $q = \begin{pmatrix} 2 \\ -1 \end{pmatrix}$, find:

a) $2p$ b) $2p - 3q$.

a) $2p = 2 \times \begin{pmatrix} 3 \\ 4 \end{pmatrix} = \begin{pmatrix} 6 \\ 8 \end{pmatrix}$

b) $2p - 3q = 2 \times \begin{pmatrix} 3 \\ 4 \end{pmatrix} - 3 \times \begin{pmatrix} 2 \\ -1 \end{pmatrix}$

$= \begin{pmatrix} 6 \\ 8 \end{pmatrix} - \begin{pmatrix} 6 \\ -3 \end{pmatrix}$

$= \begin{pmatrix} 0 \\ 11 \end{pmatrix}$

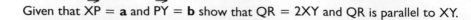

≫ Vectors in geometry

Vectors are often used to prove geometrical theorems.

Worked example

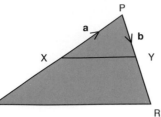

In the triangle PQR, X and Y are the midpoints of PQ and PR respectively.

Given that $\overrightarrow{XP} = \mathbf{a}$ and $\overrightarrow{PY} = \mathbf{b}$ show that QR = 2XY and QR is parallel to XY.

From the diagram:

$\overrightarrow{XY} = \overrightarrow{XP} + \overrightarrow{PY}$ so $\overrightarrow{XP} = \mathbf{a} + \mathbf{b}$

As X is the midpoint of QP then $\overrightarrow{QX} = \overrightarrow{XP}$.

Similarly as Y is the midpoint of PR then $\overrightarrow{PY} = \overrightarrow{YR}$.

From the diagram:

$\overrightarrow{QR} = \overrightarrow{QP} + \overrightarrow{PR}$

$\overrightarrow{QR} = 2\mathbf{a} + 2\mathbf{b}$ As $\overrightarrow{QP} = \overrightarrow{QX} + \overrightarrow{XP} = 2\mathbf{a}$ and $\overrightarrow{PR} = \overrightarrow{PY} + \overrightarrow{YR} = 2\mathbf{b}$.

$\overrightarrow{QR} = 2(\mathbf{a} + \mathbf{b}) = 2\overrightarrow{XY}$ As $\overrightarrow{XY} = \mathbf{a} + \mathbf{b}$ above.

This tells you that the magnitude of $\overrightarrow{QR}$ is twice the magnitude of $\overrightarrow{XY}$ so that QR = 2XY. Since $\overrightarrow{QR}$ is a multiple of $\overrightarrow{XY}$ then QR and XY are in the same direction and must, therefore be parallel.

? CHECK YOURSELF QUESTIONS

Q1 The diagram shows a series of parallel lines with $\overrightarrow{AB} = \mathbf{a}$ and $\overrightarrow{AE} = \mathbf{b}$.

Write down the following in terms of **a** and **b**.

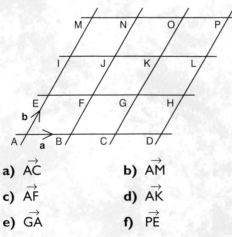

a) $\overrightarrow{AC}$ **b)** $\overrightarrow{AM}$

c) $\overrightarrow{AF}$ **d)** $\overrightarrow{AK}$

e) $\overrightarrow{GA}$ **f)** $\overrightarrow{PE}$

Q2 $\overrightarrow{AB} = \begin{pmatrix} 3 \\ 4 \end{pmatrix}$ and $\overrightarrow{BC} = \begin{pmatrix} 5 \\ -1 \end{pmatrix}$.

Find:

a) $|\overrightarrow{AB}|$

b) $|\overrightarrow{BC}|$

c) $|\overrightarrow{AC}|$.

Q3 For the regular hexagon ABCDEF with centre O, $\overrightarrow{AB} = \mathbf{a}$ and $\overrightarrow{BC} = \mathbf{b}$.

Find:
a) $\overrightarrow{AC}$ **b)** $\overrightarrow{AO}$

c) $\overrightarrow{OB}$ **d)** $\overrightarrow{AD}$.

What can you say about the quadrilateral ACDF? Give reasons for your answer.

Answers are on page 217.

UNIT 4: HANDLING DATA

Collecting and representing data

≫ How should I represent data?

Representing data is an important aspect of statistics and data handling. You must always make sure that the way you choose to represent the data is appropriate to the data you need to present. This session describes a variety of different representations with which you will need to be familiar.

≫ Classifying data

- Data that can take any values within a given range is called **continuous** data. This includes heights, temperatures, lengths and mass.

- Data that can only take particular values (such as whole or half numbers) is called **discrete** data. This includes numbers of children, separate colours and shoe sizes.

- **Quantitative** data can only take numerical values such as length, mass, capacity or temperature.

- **Qualitative** (or **categorical**) data includes qualities such as colour, taste, shade or touch.

≫ Raw data and tally charts

Raw data is information that has been collected but has not yet been organised in any way. A **tally chart** is often used to collect data. A tally chart consists of a series of tallies grouped into fives, as shown below.

Tallies	Frequency	
IIII	= 4	
ﬀ	= 5	
ﬀ I	= 6	
ﬀ II	= 7	
ﬀ ﬀ	= 10	etc.

Notice that to construct a tally chart, you make one stroke for every item of data counted, and draw every fifth stroke through the preceding four.

This makes it easy to count how many times each data item occurs. The **frequency** of an item of data is the number or times it occurs.

≫ Frequency distributions

You can easily draw a **frequency distribution** from a tally chart by totalling the tallies to find the frequencies. In some circumstances, it may be helpful to group the data and produce a grouped frequency distribution.

Worked example

The raw data gives the times (in minutes) for 20 pupils to complete a test. Construct a grouped frequency distribution using class intervals 15–17, 18–20, 21–23 and 24–26.

23	23	26	22	19	23	22	24	20	21
25	15	22	24	20	17	21	22	24	18

The grouped frequency distribution, with the given class intervals, is shown on the left.

Times	Tallies	Frequency
15–17	||	2
18–20	ЖН	4
21–23	ЖН ||||	9
24–26	||||	5

≫ Pictograms

A **pictogram** (or **pictograph** or **ideograph**) is a simple way of representing data. The frequency is indicated by a number of identical pictures. When using a pictogram, remember to include a key to explain what the individual pictures represent, as well as giving the diagram an overall title. You may also need to use a symbol that can easily be divided into halves, quarters, tenths and so on.

Worked example

Drink	Frequency
Tea	10
Coffee	13
Soup	4
Chocolate	8
Other	3

The frequency distribution on the left shows the number of different drinks purchased from a vending machine. Show this information as a pictogram.

The pictogram looks like this.

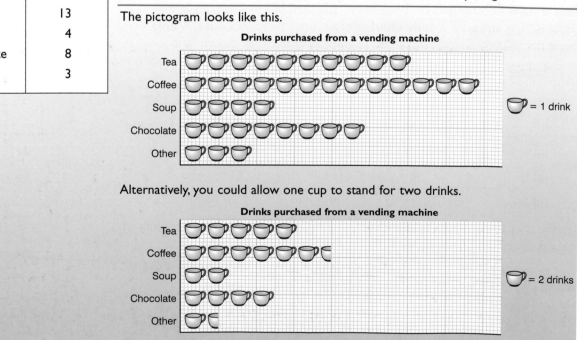

Alternatively, you could allow one cup to stand for two drinks.

≫ Stem and leaf diagrams

A **stem and leaf diagram** is a useful way of ordering raw data. When using stem and leaf diagrams, you must remember to explain the stem (with a key) and give your diagram a title, to describe what it represents.

Worked example

The daily numbers of guests in a hotel over a two-week period were recorded.

36 42 42 51 46 32 31
29 27 34 41 43 37 22

Show this information on a stem and leaf diagram. Use the tens digit as the stem.

Number of guests

stem (tens)						
2	9	7	2			
3	6	2	1	4	7	leaf (units)
4	2	2	6	1	3	
5	1					

Key: 4|2 means 42

The data may also be shown as an **ordered stem and leaf diagram**, as follows.

Number of guests

stem (tens)						
2	2	7	9			
3	1	2	4	6	7	leaf (units)
4	1	2	2	3	6	
5	1					

Key: 4|2 means 42

≫ Bar charts

A **bar chart** is a common way of representing data. The frequencies of the data items are indicated by vertical or horizontal bars, all with the same width. When using a bar chart, you must remember to label the axes clearly and give the diagram a title to explain what it represents.

Worked example

The bar chart shows the average price for cars in a number of trade magazines.

a) What was the average price for a car in the *What Car?* magazine?

b) Which magazine had the lowest average price for a car?

c) In which two magazines were the average prices the closest?

d) What is the biggest difference between average prices in the trade magazines?

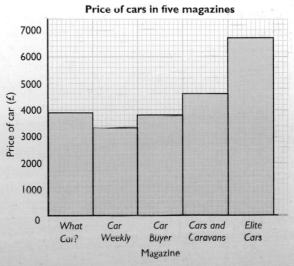

Price of cars in five magazines

From the graph:

a) The average price for a car in the *What Car?* magazine is £3900.

b) The *Car Weekly* had the lowest average price for a car.

c) The two magazines where the average prices were the closest are *What Car?* and *Car Buyer*.

d) The biggest difference between average prices is £3400 (£6700 − £3300).

≫ Line graphs

A **line graph** is another way of representing data. The frequencies are plotted at suitable points and joined by a series of straight lines. Once again, you must remember to label the axes clearly and give the diagram a title to explain what it represents.

Worked example

The information in the table shows the temperatures of a patient over a period of seven hours.

a) Draw a line graph to show this data.

b) What was the maximum recorded temperature?

c) Use your graph to find an estimate of the patient's temperature at 0930.

d) Explain why your answers in parts (b) and (c) are only approximate.

Time	Temperature (°F)
0600	102.5
0700	102.8
0800	101.5
0900	100.2
1000	99.0
1100	98.8
1200	98.6

a) This line graph shows the data.

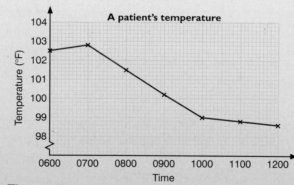

b) The maximum temperature recorded was 102.8 °F at 0700.

c) To find the patient's temperature at 0930 you read off the value at 0930.

The patient's temperature at 0930 was 99.6 °F.

d) The answers in parts (b) and (c) are only approximate as the temperature might not necessarily increase or decrease uniformly between the times at which the temperatures were taken.

≫ Frequency polygons

A **frequency polygon** can be drawn from a bar chart (or histogram), by joining the midpoints of the tops of consecutive bars, with straight lines, to form a polygon. The lines should be extended to the horizontal axis on both sides, so that the area under the frequency polygon is the same as the area under the bar chart (or histogram).

Worked example

The frequency distribution shows the heights of 50 plants, measured to the nearest centimetre.

Draw a frequency polygon to show this information.

Height (cm)	Frequency
6–10	1
11–15	3
16–20	7
21–25	9
26–30	7
31–35	10
36–40	7
41–45	5
46–50	1

The length is continuous and you can draw a bar chart (or histogram) of the information as usual. You can obtain the frequency polygon by joining up the midpoints of the tops of the bars, in order, with straight lines.

The lines should be extended to the horizontal axis on each side, as shown.

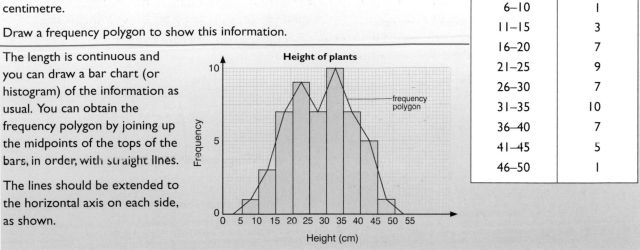

≫ Pie charts

A **pie chart** is another common way of representing data where the frequency is represented by the angles (or areas) of the sectors of a circle. When using a pie chart, you must remember to label each of the sectors clearly and give the diagram a title to explain what it represents.

The following worked examples are given to illustrate the construction of a pie chart.

Worked example

In a survey, 180 people were asked which TV channel they watched the most the previous evening. The answers to the survey are given in the table.

Construct a pie chart to show this information.

Channel	Frequency
BBC1	58
BBC2	20
ITV	42
C4	21
Channel 5	11
Other	18
Not watching	10

The pie chart needs to be drawn to represent 180 people. There are 360° in a full circle so each person will be represented by $\frac{360°}{180} = 2°$ of the pie chart.

Channel	Frequency	Angle
BBC1	58	$58 \times 2° = 116°$
BBC2	20	$20 \times 2° = 40°$
ITV	42	$42 \times 2° = 84°$
C4	21	$21 \times 2° = 42°$
Channel 5	11	$11 \times 2° = 22°$
Other	18	$18 \times 2° = 36°$
Not watching	10	$10 \times 2° = 20°$
		$\overline{360°}$

TV channels watched

HANDLING DATA

Q1 The table shows the sales of different coloured socks in a department store.

Colour	Frequency
White	18
Black	13
Blue	10
Brown	6
Other	7

Show this information as:

a) a pictogram   **b)** a bar chart.

Q2 The following information shows the heights of 12 plants, in centimetres.

4.9   3.3   4.0   5.0   3.8   4.6
3.2   4.2   5.1   5.1   3.7   4.9

Show this information as an ordered stem and leaf diagram.

Q3 The weight of a child is recorded at birth and at the end of each month as shown in the table.

Age (months)	0	1	2	3	4	5
Weight (pounds)	8	9.5	10.8	12.4	13.8	15.2

Draw a line graph to represent this information.

Q4 The lengths of 100 bolts are measured and recorded in this table.

Length (mm)	20–25	25–30	30–35	35–40	40–45
Frequency	3	8	15	7	2

Draw a frequency polygon to show this information.

Q5 250 students at a college were asked about the courses they are following. Their responses were as follows.

A levels   106
GCSEs   42
GNVQs   86
Others   16

Construct the pie chart to show the different courses.

Q6 This pie chart shows how 100 stockbrokers travelled to work one day.

a) Which method of transport was the most popular?

b) Which method of transport was the least popular?

c) What angle is represented by the 'cycle' sector?

Twice as many travelled to work by bus as by car.

d) How many stockbrokers travelled to work by bus?

e) How many stockbrokers travelled to work by car?

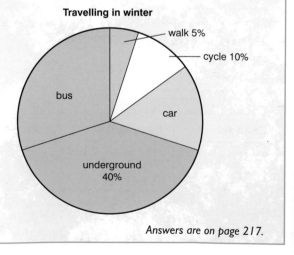

Travelling in winter

walk 5%
cycle 10%
bus
car
underground 40%

Answers are on page 217.

≫ What does it mean?

Measures of central tendency are more often referred to as **measures of average**. You will need to understand the differences between the **mode**, **median** and **mean**, and be able to find them, for the examination.

≫ Mode of a distribution

The **mode** of a distribution is the value that occurs most frequently. If there are two modes then the distribution is called **bimodal**. If there are more than two modes then the distribution is called **multimodal**.

Worked example
Find the mode of the following distribution.

8, 6, 7, 4, 9, 8, 8, 6, 7, 6, 8

The number 4 occurs 1 time.

The number 5 occurs 0 times.

The number 6 occurs 3 times.

The number 7 occurs 2 times.

The number 8 occurs 4 times.

The number 9 occurs 1 time.

The number 8 occurs the most frequently so the mode is 8.

+ *HINT*

▸ Make sure that you write down the *value* of the mode and not the *frequency*.

MODE OF A FREQUENCY DISTRIBUTION
The mode of frequency distribution is the value that has the highest frequency.

Worked example
Find the mode of this frequency distribution.

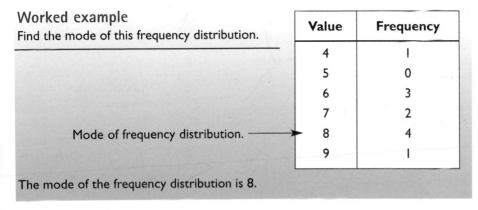

Mode of frequency distribution. ——→

Value	Frequency
4	1
5	0
6	3
7	2
8	4
9	1

The mode of the frequency distribution is **8**.

MODE OF A GROUPED FREQUENCY DISTRIBUTION

The mode of a grouped frequency distribution has little meaning, although it is possible to identify a **modal group**.

Worked example

Find the modal group of this grouped frequency distribution.

Weight (grams)	Frequency
15–25	11
25–35	17
35–45	23
45–55	16
55–65	10

Modal group. ➝ 35–45

The modal group of the grouped frequency distribution is 35–45.

≫ Median of a distribution

The **median** of a distribution is the middle value when the values are arranged in order. Where there are two middle values (i.e. for an even number of values) then you add the two numbers and divide by 2.

+ HINT

▸ If there are n values in the distribution then the median position is given by $\frac{1}{2}(n + 1)$.

Worked example

Find the median of each of the following distributions.

a) 8, 6, 7, 4, 9, 8, 8, 6, 7, 6, 8

b) 8, 6, 7, 4, 9, 8, 8, 6, 7, 6, 8, 10

a) For the distribution: 8, 6, 7, 4, 9, 8, 8, 6, 7, 6, 8

Rearrange in order: 4, 6, 6, 6, 7, 7, 8, 8, 8, 8, 9

The median position is given by $\frac{1}{2}(n + 1) = \frac{1}{2}(11 + 1) = $ 6th value.

4, 6, 6, 6, 7, ⑦, 8, 8, 8, 8, 9

So the median is 7.

b) For the distribution: 8, 6, 7, 4, 9, 8, 8, 6, 7, 6, 8, 10

Rearrange in order: 4, 6, 6, 6, 7, 7, 8, 8, 8, 8, 9, 10

The median position is given by $\frac{1}{2}(n + 1) = \frac{1}{2}(12 + 1)$

$$= 6\frac{1}{2}\text{th value}$$

(i.e. between the 6th and 7th values).

4, 6, 6, 6, 7, ⑦, ⑧, 8, 8, 8, 9, 10

So the median is $\frac{1}{2}(7 + 8) = 7\frac{1}{2}$.

MEDIAN OF A FREQUENCY DISTRIBUTION

To find the median of a frequency distribution you need to work out the **cumulative frequency distribution**, as shown in the next example.

Worked example

Find the median of the frequency distribution in this table on the right.

Value	Frequency
4	1
5	0
6	3
7	2
8	4
9	1

To find the median of a frequency distribution you can work out the cumulative frequency like this.

From the cumulative frequency you can see that the middle value (i.e. the 6th value) occurs at 7.

Value	Frequency	Cumulative frequency
4	1	1
5	0	1
6	3	4
7	2	6
8	4	10
9	1	11

To find the median of a group frequency distribution you need to draw up a cumulative frequency diagram. Cumulative frequency diagrams are considered fully in Revision session 4, *Measures of spread and cumulative frequency diagrams*.

≫ Mean of a distribution

The **mean** (or **arithmetic mean**) of a distribution is found by summing the values of the distribution and dividing by the number of values.

Worked example

Find the mean of the following distribution.

8, 6, 7, 4, 9, 8, 8, 6, 7, 6, 8

Add the values of the distribution and dividing by the number of values.

$$\text{Mean} = \frac{8 + 6 + 7 + 4 + 9 + 8 + 8 + 6 + 7 + 6 + 8}{11} = \frac{77}{11} = 7$$

The definition for the mean is often written as:

$$\text{mean} = \frac{\text{sum of the values}}{\text{number of values}} \quad \text{or} \quad \text{mean} = \frac{\Sigma \text{ values}}{\text{number of values}}$$

MEAN OF A FREQUENCY DISTRIBUTION

You can find the mean of a frequency distribution by summing the values of the distribution and dividing by the number of values.

$$\text{Mean} = \frac{\text{sum of the values}}{\text{number of values}}$$

For a frequency distribution the sum of the values is equal to the sum of the products (frequency × value) or Σfx.

The number of values is the sum of the frequencies Σf. So the mean is $\frac{\Sigma fx}{\Sigma f}$.

+ HINT

▶ Note that Σ means 'the sum or total of'.

Worked example

Find the mean of this frequency distribution.

Value x	Frequency f	Frequency × value fx
4	1	1 × 4 = 4
5	0	0 × 5 = 0
6	3	3 × 6 = 18
7	2	2 × 7 = 14
8	4	4 × 8 = 32
9	1	1 × 9 = 9
	$\Sigma f = 11$	$\Sigma fx = 77$

Mean for the frequency distribution $= \dfrac{\Sigma fx}{\Sigma f} = \dfrac{77}{11} = 7$ (as before)

＋ HINTS

▶ The mid-interval values are used as an estimate of the particular interval so that the final answer will not be exact but will be an 'estimate of the mean'.

▶ The mid-interval value is found by taking the mean of the upper and lower class boundaries – See Revision session 3, *Measures of spread*, for definitions.

≫ Mean of a grouped frequency distribution

You can find the mean of a grouped frequency distribution in the same way as for a frequency distribution, using the **mid-interval values** (or midpoints) as representative of the interval.

Worked example

The following table shows the heights of trees growing in a nursery. Calculate an estimate of the mean height of the trees.

Height (cm)	15–20	20–30	30–40	40–50	50–60	60–70	70–80
Frequency	8	4	5	11	17	2	1

Height	Mid-interval value x	Frequency f	Frequency × mid-interval value fx
15–20	17.5	8	8 × 17.5 = 140
20–30	25	4	4 × 25 = 100
30–40	35	5	5 × 35 = 175
40–50	45	11	11 × 45 = 495
50–60	55	17	17 × 55 = 935
60–70	65	2	2 × 65 = 130
70–80	75	1	1 × 75 = 75
		$\Sigma f = 48$	$\Sigma fx = 2050$

For the grouped frequency distribution:

mean $= \dfrac{\Sigma fx}{\Sigma f} = \dfrac{2050}{48} = 42.708\,333\,3 = 43\,\text{cm}$ to an appropriate degree of accuracy.

An answer of 43 cm is appropriate, bearing in mind the accuracy of the original data and the inaccuracies resulting from the use of the mid-interval values as an estimate of the particular interval.

CHECK YOURSELF QUESTIONS

Q1 Find the mode of the following pictorial representations.

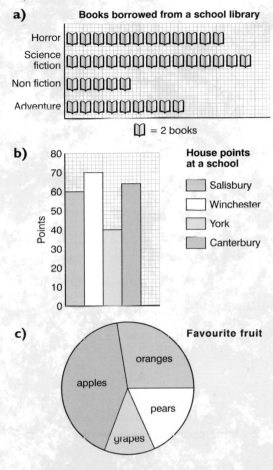

a) Books borrowed from a school library

Horror
Science fiction
Non fiction
Adventure

📖 = 2 books

b) House points at a school

Points

- ▨ Salisbury
- ☐ Winchester
- ▨ York
- ▨ Canterbury

c) Favourite fruit

oranges
apples
pears
grapes

Q2 Find the median of the following frequency distribution which shows the number of goals scored in **34** premier division matches.

Number of goals	Frequency
0	4
1	8
2	11
3	7
4	3
5	0
6	1

Q3 Find the mean of the following data.

Age (years)	17	18	19	20	21
Frequency	23	13	4	0	1

Q4 Calculate an estimate of the mean for this distribution.

Weight (kg)	0–10	10–20	20–30	30–40	40–50	50–60
Frequency	11	18	16	11	5	2

Q5 The amount of money spent by customers in a restaurant is shown in this table.

Use the information to calculate an estimate of the mean.

Amount (£)	Frequency
0 and less than 5	12
5 and less than 10	15
10 and less than 15	8
15 and less than 20	7
20 and less than 25	3

Answers are on page 218.

Measures of spread and cumulative frequency diagrams

≫ What are measures of spread?

The measure of spread gives a measure of how spread out the data values are. For the examination you need to be familiar with the **range** and **interquartile range**.

≫ The range

You can find the **range** of a distribution by working out the difference between the greatest value and least value. You should always give the range as a single value.

Worked example

The insurance premiums paid by eleven households are listed below.

£340   £355   £400   £320   £380   £320   £632   £365   £340   £380   £370

Calculate the mean and the range.

$$\text{Mean} = \frac{£340 + £355 + £400 + £320 + £380 + £320 + £632 + £365 + £340 + £380 + £370}{11}$$

$$\text{Mean} = \frac{4202}{11}$$

$$= £382$$

The mean of £382 is deceptive as a measure of central tendency because it is affected by the value of £632 – this type of value is sometimes called an **extreme value**.

Greatest value = £632

Least value = £320

Range = greatest value − least value

$$= £632 − £320$$

$$= £312$$

Similarly, the range of £312 is deceptive as a measure of spread because again, it is also affected by the value of £632.

≫ Interquartile range

Although the range is affected by extreme values, the **interquartile range** considers only the middle 50% of the distribution.

You can find the interquartile range by dividing the data into four parts or **quartiles** and working out the difference between the upper quartile and the lower quartile, as shown in the following worked example.

Worked example

The insurance premiums paid by eleven households are listed below.

£340   £355   £400   £320   £380   £320   £632   £365   £340   £380   £370

Calculate the interquartile range.

Arranging the data in order and considering the middle 50%:

£320   £320   £340   £340   £355   £365   £370   £380   £380   £400   £632
                  ↑   ↑   ↑
                  LQ   Median   UQ

Upper quartile = £380

Lower quartile = £340

Interquartile range = upper quartile − lower quartile
                    = £380 − £340
                    = £40

You have already seen that if there are n values in the distribution then the median position is given by $\frac{1}{2}(n + 1)$.

Similarly, the lower quartile position is given by $\frac{1}{4}(n + 1)$ and the upper quartile position is given by $\frac{3}{4}(n + 1)$.

≫ Box and whisker plots

Another useful way of showing a frequency distribution is a **box and whisker plot** (sometimes called a **box plot**). The plot shows:

- the **median**
- the **upper and lower quartiles**
- the **maximum and minimum values.**

Worked example

For the data about the insurance premiums in the previous examples, the median is £365, the upper quartile is £380, the lower quartile is £340, the maximum value is £632 and the minimum value is £320. Draw the box plot.

The box plot looks like this.

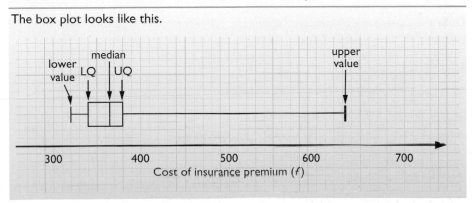

≫ Cumulative frequency diagrams

Cumulative frequency diagrams can be used to find the median and the quartiles of a variety of distributions, including grouped frequency distributions. To find the cumulative frequency you find the accumulated totals and plot them on the cumulative frequency diagram (or **ogive**), then join them with a smooth curve.

CLASS LIMITS

Class limits are the values given in each of the individual groups (or **class intervals**). For the class interval 1–3, in the example below, the class limits are 1 and 3 (where the **lower class limit** is 1 and the **upper class limit** is 3).

CLASS BOUNDARIES

As the times are given to the nearest minute then the interval 1–3 will actually include times from 0.5 minutes to 3.5 minutes. The **class boundaries** are 0.5 and 3.5 (where the **lower class boundary** is 0.5 and the **upper class boundary** is 3.5).

CLASS WIDTH

The **class width, class length** or **class size** is the difference between the upper and lower class boundaries.

For the class interval 1–3 with lower class boundary 0.5 and upper class boundary 3.5, then the class width equals 3.5 – 0.5 = 3 minutes.

Worked example

The table shows the times (given to the nearest minute) that customers have to wait in a checkout queue.

Waiting time (minutes)	Frequency
1–3	8
4–6	19
7–9	11
10–12	6
13–15	2
16–18	1

Draw the cumulative frequency diagram for the information. Use your diagram to find:

a) how many customers waited less than 5 minutes

b) how many waited more than 10 minutes

c) the median and the interquartile range.

First complete the table to include the cumulative frequencies.
Then draw the cumulative frequency diagram.

Waiting time (minutes)	Frequency	Cumulative frequency
1–3	8	8
4–6	19	27
7–9	11	38
10–12	6	44
13–15	2	46
16–18	1	47

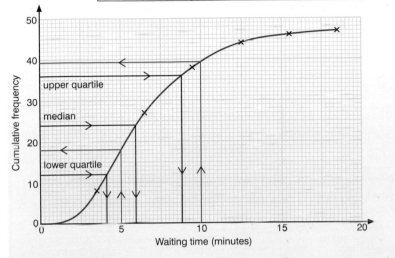

HINTS

▸ The final cumulative frequency should equal the sum of the frequencies.
▸ The cumulative frequencies must be plotted at the upper class boundaries (i.e. 3.5, 6.5, 9.5, 12.5, 15.5 and 18.5).

a) To find out how many customers waited less than 5 minutes, read the information against a waiting time of 5 minutes.

From the graph, the number of customers who waited less than 5 minutes is 18.

b) To find out how many customers waited more than 10 minutes, read the information against a waiting time of 10 minutes.

From the graph, the number of customers who waited less than 10 minutes is 39 so that the number of customers who waited more than 10 minutes is $47 - 39 = 8$.

c) To find the median waiting time you need to read off the median value on the cumulative frequency.

The median position is given by $\frac{1}{2}(n + 1) = \frac{1}{2}(47 + 1) = $ 24th value.

To find the interquartile range you need to find the lower quartile and the upper quartile.

The lower quartile position is given by $\frac{1}{4}(n + 1) = \frac{1}{4}(47 + 1) = $ 12th value.

The upper quartile position is given by $\frac{3}{4}(n + 1) = \frac{3}{4}(47 + 1) = $ 36th value.

From the graph, the median = 6.

Similarly, from the graph, the upper quartile = 8.8 and
the lower quartile = 4.1

so the interquartile range = upper quartile – lower quartile
$$= 8.8 - 4.1$$
$$= 4.7$$

Q1 The following information gives the heights of 15 plants.

13 cm, 36 cm, 15 cm, 13 cm, 21 cm, 18 cm, 37 cm, 18 cm, 12 cm, 21 cm, 18 cm, 20 cm, 6 cm, 37 cm, 39 cm

Find the range and interquartile range.

Q2 Use the box and whisker plot below to calculate:

a) the range

b) the interquartile range of the data.

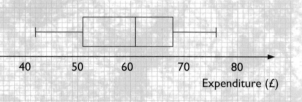

Expenditure (£)

Q3 The frequency distribution for the time taken to obtain clearance through customs is given in the following table.

Time (t minutes)	Frequency
$20 \leqslant t < 25$	3
$25 \leqslant t < 30$	7
$30 \leqslant t < 35$	15
$35 \leqslant t < 40$	18
$40 \leqslant t < 45$	22
$45 \leqslant t < 50$	17
$50 \leqslant t < 55$	8
$55 \leqslant t < 60$	2

Draw a cumulative frequency curve and use it to find an estimate of the median and the interquartile range.

Q4 The following table shows the number of words per paragraph in a children's book.

Number of words per paragraph	Number of paragraphs
1–10	17
11–20	33
21–30	51
31–40	21
41–50	18

Draw a cumulative frequency curve to illustrate this information and use your graph to estimate:

a) the median and interquartile range

b) the percentage of paragraphs over 35 words in length.

Answers are on page 219.

Scatter diagrams and lines of best fit

≫ What are scatter diagrams?

Scatter diagrams are used to show the relationship between two variables. Each of the two variables is assigned to a different axis and the information is plotted as coordinates on the scatter diagram.

≫ Using scatter diagrams

Worked example

The following table shows the heights and shoe sizes of 10 pupils.

Shoe size	3	2	5	$6\frac{1}{2}$	4	3	6	1	$3\frac{1}{2}$	$7\frac{1}{2}$
Height (cm)	133	126	150	158	135	128	152	118	142	101

Draw a scatter diagram of the information.

By considering each pair of values as a different pair of coordinates you can produce this scatter diagram.

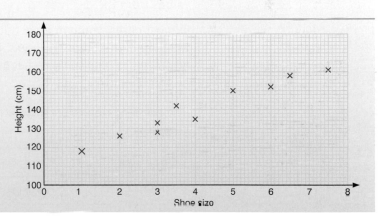

≫ Correlation

You can use scatter graphs to show whether there is any **relationship** or **correlation** between two variables. You need to know the following descriptions of such relationships or correlation.

Little or no apparent correlation

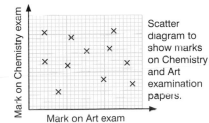

Scatter diagram to show marks on Chemistry and Art examination papers.

The points are scattered randomly over the graph, indicating little or no correlation between the two variables.

Moderate correlation

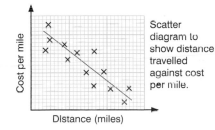

Scatter diagram to show distance travelled against cost per mile.

The points lie close to a straight line, indicating moderate correlation between the two variables (the closer the points are to the line then the stronger the correlation).

Strong correlation

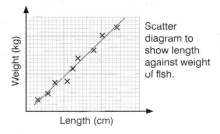

Scatter diagram to show length against weight of fish.

The points lie along a straight line, indicating a strong correlation between the two variables.

Correlation can also be described as **positive** or **negative**.

Positive correlation

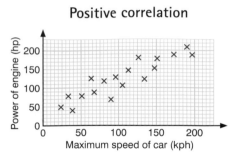

Negative correlation

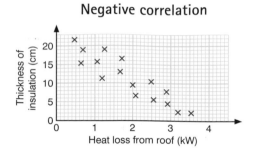

Where an increase in one variable is associated with an increase in the other variable then the correlation is said to be positive or **direct**.

Where an increase in one variable is associated with a decrease in the other variable then the correlation is said to be negative or **inverse**.

≫ Line of best fit

Where the points on a scatter diagram show moderate or strong correlation you can draw a line to approximate the relationship. This line is called the **line of best fit** (or **regression line**) and it can be used to predict other values from the given data.

In most cases, you can draw a line of best fit 'by eye'. For more accurate work the line of best fit should pass through $(\overline{x}, \overline{y})$ where $\overline{x}$ and $\overline{y}$ are the mean values of x and y respectively.

Worked example

The following table (from the previous worked example) shows the heights and shoe sizes of 10 pupils.

Shoe size	3	2	5	$6\frac{1}{2}$	4	3	6	1	$3\frac{1}{2}$	$7\frac{1}{2}$
Height (cm)	133	126	150	158	135	128	152	118	142	161

Draw a scatter diagram and the line of best fit.

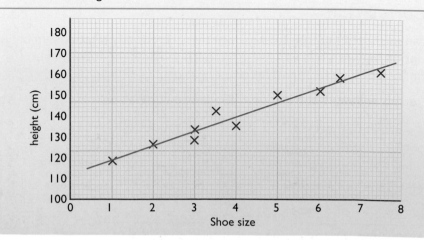

? CHECK YOURSELF QUESTIONS

Q1 Describe fully the relationship between the two variables in each of the following scatter diagrams.

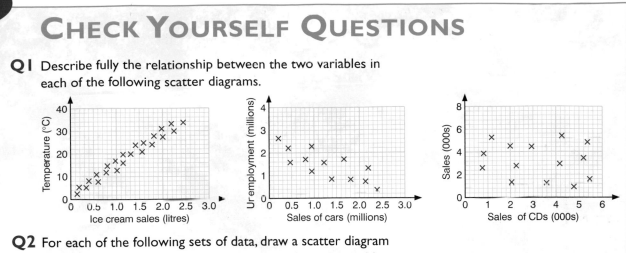

Q2 For each of the following sets of data, draw a scatter diagram and describe fully the relationship between the two variables.

a)

Variable P	19	17	5	14	26	5	24	8
Variable Q	13	19	37	21	4	33	10	29

b)

Variable M	30	10	20	37	16	45	55	25	46
Variable N	68	20	57	48	40	47	21	57	30

Q3 The following information shows the temperature and the production rate for a factory.

Temperature (°C)	23.4	21.4	25.1	23.4	24.8	21.7	20.9	22.7	21.7	23.8	22.2
Production rate (units)	47	53	35	53	41	58	60	59	64	40	50

Draw a scatter diagram and show the line of best fit.

Use your line of best fit to calculate:

a) the production rate for a temperature of 22.8 °C

b) the temperature for a production rate of 50 units.

Q4 The following information shows the marks awarded to students on two examination papers.

Paper 1	30	56	40	68	14	85	64	28	79	48	44	59
Paper 2	30	39	20	73	16	83	45	15	62	44	32	61

Draw the line of best fit and use it to calculate:

a) the mark of someone who gets 65 on paper 1

b) the mark of someone who gets 75 on paper 2.

Answers are on page 220.

≫ What is probability?

Probability is the branch of statistics that allows you to work out how likely or unlikely an **outcome** or result of an **event** might be.

In probability, an outcome that is certain to happen has a probability of 1 and an event that is impossible has a probability of 0. Probabilities greater than 1 or less than 0 have no meaning.

Theoretical probability is based on equally likely outcomes. You can use it to tell how an event should perform in theory, whereas **experimental probability** (or **relative frequency**) tells you how an event performs in an experiment.

≫ Events and outcomes

An **event** is something that happens, such as throwing a die, or tossing a coin, or picking a card from a pack.

An **outcome** is the result of an event, such as scoring a 3 or a 6 when you throw a die.

If the outcome is the required result, such as throwing a 6 to start a game, then the outcome is a **success**.

The **probability** is a measure of how likely an outcome is to happen.

In general:

$$\text{probability of success} = \frac{\text{number of 'successful' outcomes}}{\text{number of 'possible' outcomes}}$$

and you can use p(success) as shorthand for the probability of success.

$$p(\text{success}) = \frac{\text{number of 'successful' outcomes}}{\text{number of 'possible' outcomes}}$$

Worked example

A box contains 25 coloured balls, where seven balls are red, ten balls are blue and eight balls are yellow. A ball is selected from the box at random. Calculate the probability of selecting:

a) a red ball **b)** a blue ball **c)** a red or a yellow ball

d) a red or a blue or a yellow ball **e)** a green ball.

Use:

$$p(\text{success}) = \frac{\text{number of 'successful' outcomes}}{\text{number of 'possible' outcomes}}$$

a) $p(\text{red ball}) = \dfrac{\text{number of red balls}}{\text{number of balls}} = \dfrac{7}{25}$

b) $p(\text{blue ball}) = \dfrac{\text{number of blue balls}}{\text{number of balls}} = \dfrac{10}{25} = \dfrac{2}{5}$ Cancelling to lowest terms.

+ HINT

▸ You could also give the answer as a decimal or a percentage.
So p(red ball) = 0.28
or p(red ball) = 28%

c) $p(\text{red or yellow ball}) = \dfrac{\text{number of red or yellow balls}}{\text{number of balls}}$

$$= \frac{15}{25} = \frac{3}{5} \quad \text{Cancelling to lowest terms.}$$

d) $p(\text{red or yellow or blue ball}) = \dfrac{\text{number of red or yellow or blue balls}}{\text{number of balls}}$

$$= \frac{25}{25}$$

$$= 1 \quad \text{Meaning that this outcome is certain to happen.}$$

e) $p(\text{green ball}) = \dfrac{\text{number of green balls}}{\text{number of balls}} = \dfrac{0}{25}$

$$= 0 \quad \text{Meaning that this outcome is impossible, it cannot happen.}$$

TOTAL PROBABILITY

The probability of an outcome happening is equal to 1 minus the probability of the outcome not happening.

$$p(\text{outcome occurs}) = 1 - p(\text{outcome does not occur})$$

Worked example

The probability that it will rain tomorrow is $\frac{1}{5}$. What is the probability that it will not rain tomorrow?

$p(\text{rain}) = \frac{1}{5}$

$p(\text{not rain}) = 1 - \frac{1}{5} = \frac{4}{5}$ $\qquad p(\text{not rain}) = 1 - p(\text{rain})$

≫ Possibility spaces

A **possibility space** is a diagram which can be used to show the outcomes of various events.

Worked example

Two fair dice are thrown and the sum of the scores on the faces is noted. What is the probability that the sum is 8?

Draw a diagram to illustrate the possible outcomes.

The diagram to illustrate the possible outcomes is shown below. There are 36 possible outcomes. There are 5 outcomes that give a total of 8.

Probability that the sum is 8 $= \frac{5}{36}$.

		Second die					
		1	2	3	4	5	6
	1	2	3	4	5	6	7
	2	3	4	5	6	7	8
First	3	4	5	6	7	8	9
die	4	5	6	7	8	9	10
	5	6	7	8	9	10	11
	6	7	8	9	10	11	12

+ HINT

▸ You may see the word 'dice' used instead of 'die' in these questions.
Die is the singular form, dice is the plural.

HANDLING DATA

Worked example

A die is thrown 100 times.

a) How many times would you expect to throw a six?

This frequency distribution is obtained.

b) What is the relative frequency of a score of 6?

Score	1	2	3	4	5	6
Frequency	18	15	19	17	16	15

c) What is the relative frequency of getting an even number?

d) For which score are the theoretical probability and relative frequency the closest?

a) When throwing a die 100 times:

expected number of sixes $= 100 \times \frac{1}{6} = 16.66666\ldots$

$\qquad\qquad\qquad\qquad = 17$ (to the nearest whole number)

b) The relative frequency of a score of 6 is $\frac{15}{100} = \frac{3}{20}$.

c) The frequency of getting an even number is $15 + 17 + 15 = 47$.

The relative frequency of getting an even number is $\frac{47}{100}$.

d) The theoretical and experimental probabilities are closest for a score of 4.

? CHECK YOURSELF QUESTIONS

Q1 A box contains 50 balls coloured blue, red and green. The probability of getting a blue ball is 32% and the probability of getting a red ball is 0.46.

    a) How many blue balls are there in the box?

    b) How many red balls are there in the box?

    c) How many green balls are there in the box?

Q2 The probability that a train arrives early is 0.2 and the probability that the train arrives on time is 0.45. What is the probability that the train arrives late?

Q3 An experiment consists of throwing a die and tossing a coin. Draw a possibility space for the two events and use this to calculate the probability of scoring:

    a) a head and a 1

    b) a tail and an odd number.

Q4 Two tetrahedral dice, each numbered 1 to 4, are thrown simultaneously. Draw a possibility space for the total of the two dice and use this information to calculate the probability of scoring a total of:

    a) 2   **b)** 6   **c)** 9.

What is the most likely outcome?

Q5 A die is thrown 120 times. What is the expected frequency of a number greater than 4?

Q6 The probability that a new car will develop a fault in the first month after delivery is 0.062%. A garage sells 1037 new cars in one year. How many of these cars will be expected to develop a fault in the first month after delivery?

Answers are on page 221.

The addition and multiplication rules

≫ What definitions do I need to know?

Two or more events are **mutually exclusive** if they cannot happen at the same time. Two events are **independent** if the outcome of one does not affect the outcome of the other.

≫ Mutually exclusive events

THE ADDITION RULE

When you are working with **mutually exclusive** events you can apply the **addition rule** (also called the **or rule**) which states that:

$p(A \text{ or } B) = p(A) + p(B)$

and for more than two mutually exclusive events:

$p(A \text{ or } B \text{ or } C \text{ or } ...) = p(A) + p(B) + p(C) + ...$.

Worked example

A spinner with ten sides, numbered 1 to 10, is spun. What is the probability of getting:

a) a five

b) a five or a six

c) a multiple of 3 or a multiple of 4

d) a multiple of 2 or a multiple of 3?

a) $p(5) = \frac{1}{10}$

b) $p(5 \text{ or } 6) = p(5) + p(6)$ As the events are mutually exclusive

$= \frac{1}{10} + \frac{1}{10}$

$= \frac{2}{10}$

$= \frac{1}{5}$ Cancelling down to the lowest terms.

c) $p(\text{multiple of 3 or multiple of 4})$
$= p(3) + p(6) + p(9) + p(4) + p(8)$ As the events are mutually exclusive.

$= \frac{1}{10} + \frac{1}{10} + \frac{1}{10} + \frac{1}{10} + \frac{1}{10} = \frac{5}{10}$

$= \frac{1}{2}$ Cancelling down to the lowest terms.

d) $p(\text{multiple of 2 or multiple of 3})$
$= p(2 \text{ or } 3 \text{ or } 4 \text{ or } 6 \text{ or } 8 \text{ or } 9 \text{ or } 10)$

As the events are not mutually exclusive, because the number 6 is common to both events and if the probabilities are added then this probability will be added twice.

$= p(2) + p(3) + p(4) + p(6) + p(8) + p(9) + p(10)$
As these events are now mutually exclusive.

$= \frac{1}{10} + \frac{1}{10} + \frac{1}{10} + \frac{1}{10} + \frac{1}{10} + \frac{1}{10} + \frac{1}{10}$

$= \frac{7}{10}$

≫ Independent events

In a **tree diagram**, you write the probabilities of the outcomes of different events on different branches of the 'tree'.

Worked example

A bag contains four red and three blue counters. A counter is drawn from the bag, replaced and then a second counter is drawn from the bag. Draw a tree diagram to show the various possibilities that can occur.

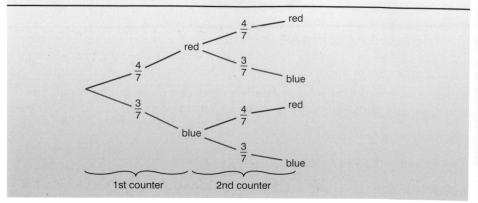

THE MULTIPLICATION RULE

For **independent events** you can use the **multiplication rule** (also called the **and rule**) which states that:

$$p(A \text{ and } B) = p(A) \times p(B)$$

Similarly for more than two independent events:

$$p(A \text{ and } B \text{ and } C \text{ and } ...) = p(A) \times p(B) \times p(C) \times ...$$

Worked example

A bag contains four red and three blue counters. A counter is drawn from the bag, replaced and then a second counter is drawn from the bag. Draw a tree diagram and use it to calculate the probability that:

a) both counters will be red   **b)** both counters will be blue

c) the first counter will be red and the second counter blue

d) one counter will be red and one counter will be blue.

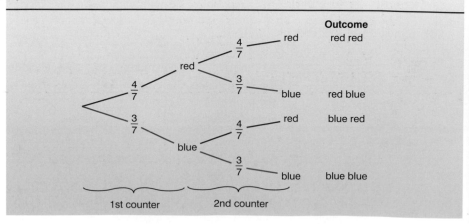

a) p(red and red) = p(red) × p(red)   As the events are independent.

$$= \frac{4}{7} \times \frac{4}{7} = \frac{16}{49}$$

b) p(blue and blue) = p(blue) × p(blue)   As the events are independent.

$$= \frac{3}{7} \times \frac{3}{7} = \frac{9}{49}$$

c) p(red and blue) = p(red) × p(blue)   As the events are independent.

$$= \frac{4}{7} \times \frac{3}{7} = \frac{12}{49}$$

d) p(one counter will be red and one counter will be blue) is the same as p(red and blue or blue and red).

= p (red and blue) + p(blue and red). As the events are mutually exclusive.

= p(red) × p(blue) + p(blue) × p(red) As the events are independent.

$$= \frac{4}{7} \times \frac{3}{7} + \frac{3}{7} \times \frac{4}{7} = \frac{12}{49} + \frac{12}{49} = \frac{24}{49}$$

Both of these outcomes give one red and one blue.

? CHECK YOURSELF QUESTIONS

Q1 A die with faces numbered 1 to 6 is rolled and the value on the face uppermost is noted. Find the probability that the result will be:

a) a 5 or a 6   **b)** an even number

c) a factor of 8.

Q2 Letters are chosen from the word:

PROBABILITY

Find the probability that the chosen letter is:

a) the letter P   **b)** the letter B

c) the letter B or the letter I.

Q3 A counter is selected from a box containing three red, four green and five blue counters and a second counter is selected from a different box containing five red and four green counters. Draw a tree diagram to show the various possibilities when a counter is drawn from each bag.

Q4 The probability that a car will fail its MOT because of the lights is 0.32 and the probability that a car will fail its MOT because of the brakes is 0.55. Calculate the probability that the car fails because of:

a) its lights and its brakes

b) its lights only.

Q5 The probability that a particular component will fail is 0.015. Draw and label a tree diagram to show the possible outcomes when two such components are chosen at random. Calculate the probability that:

a) both components will fail

b) exactly one component will fail.

Answers are on page 222.

Sampling methods

HINT

▶ You may find that some of these words are useful for your coursework.

≫ Types of data

PRIMARY AND SECONDARY DATA

Primary data is collected as part of a statistical investigation such as a census or survey, whereas **secondary** data is data which already exists. Once primary data is collected and processed then it becomes secondary data. Examples of secondary data include information provided by government departments, businesses, market research companies. Data may be collected by:

- observation – collecting information and using some means of recording it, such as observation sheets, tape recorders or video recorders. This form of data collection can include **systematic** observation, where the observant tries to be as unobtrusive as possible, or else **participant** observation, where the observer participates in the activity.

- interviewing – asking questions of individuals, or groups of individuals, using some set format. Interviewing can be **formal**, where the questions will follow a strict format, or else **informal**, where the questions will follow some general format.

- questionnaires – the most popular method of collecting data. It usually involves postal questionnaires or esle questionnaires that are left for the respondents to complete in their own time. A good questionnaire should be simple, short, clear and precise and the questions should be unambiguous, written in appropriate language, avoid personal or offensive questions and be free from bias.

PILOT SURVEY

The pilot survey is a preliminary survey carried out on a small number of people. The pilot survey is useful to check for likely problems and highlight areas requiring further clarification before the actual survey is undertaken.

≫ What is a sample?

A **sample** is a selection of data, taken from a larger population. It should be representative of the population as a whole and sample sizes should be as large as possible. Sampling techniques are described below.

CONVENIENCE SAMPLING

In **convenience sampling** or **opportunity sampling,** you just choose the first people who come along. This might mean sampling friends and members of your own family and is therefore likely to involve some element of bias.

RANDOM SAMPLING

In **random sampling**, each member of the population has an equally likely chance of being selected. Random sampling might involve giving each member of the population a number and then choosing the numbers at random, using some appropriate means to generate random numbers.

SYSTEMATIC SAMPLING

Systematic sampling involves random sampling, using some system to choose the members of the population to be sampled. Systematic sampling might include numbering each member of the population according to some criterion, such as name, age, height.

STRATIFIED SAMPLING

Stratified sampling involves dividing the population into groups or **strata**. From each stratum you choose a random or systematic sample so that the sample size is proportional to the size of the group in the population as a whole. For example, in a class where there are twice as many girls as boys, the sample should include twice as many girls as boys.

QUOTA SAMPLING

Quota sampling involves choosing population members who have specific characteristics that are selected beforehand. Quota sampling is popular in market research where interviewers are told, for example, that there should be equal numbers of men and women, or twice as many adults as teenagers, or that the sample should include ten shoppers and five commuters.

CHECK YOURSELF QUESTIONS

Q1 A bus company attempted to estimate the number of people who travel on local buses in a certain town. They telephoned 100 people in the town one evening and asked, 'Have you travelled by bus in the last week?'. Nineteen people said 'Yes'.

The bus company concluded that 19% of the town's population travel on local buses. Give three criticisms of this method of estimation.

Q2 The numbers of people living in three villages are given in the table.

Village	Population
Atford	2500
Beeham	4100
Calbridge	5900

A sample of 240 is taken by Mr James. He selects, at random, 80 people from each village.

a) Explain why this might be an inappropriate sampling method.

b) Explain how Mr James could select a more representative sample of 240 people, and write down the number of people selected from each village.

Q3 Give two advantages of using a pilot survey.

Q4 The following three questions were found on a questionnaire.

a) 'What do you think of our improved magazine?'

b) 'How many hours of television do you watch?'

c) 'Do you or do you not listen to the radio?'

What is wrong with these questions?

Answers are on page 223.

Histograms – unequal class intervals

≫ What are histograms?

Histograms are like bar charts except that it is the *area* of each bar that represents the frequency, rather than the *length* or *height*.

≫ Using histograms

You should draw the bars on the horizontal axis at the class boundaries and the area of the bars should be proportional to the frequency, i.e.

class width × height = frequency

so height = $\dfrac{\text{frequency}}{\text{class width}}$

and the height is referred to as the **frequency density**. This means that the vertical axis of a histogram should be labelled frequency density where:

frequency density = $\dfrac{\text{frequency}}{\text{class width}}$

Worked example

The histogram shows the ages of cars in an office car park.

a) How many cars in the sample were from two to four years old?

b) How many cars were there in the sample altogether?

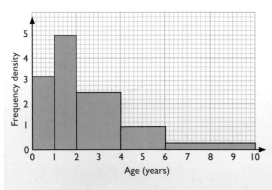

Remember that the frequency is represented by the area and is found by multiplying the frequency density (on the vertical axis) by the class width (on the horizontal axis).

a) The number of cars that are from two to four years old
= frequency density × class width
= 2.5 × 2
= 5

b) The total number of cars can be calculated using the above method for each class width.

Age (years)	0–1	1–2	2–4	4–6	6–10
Frequency density	3	5	2.5	1	0.25
Frequency	3 × 1 = 3	5 × 1 = 5	2.5 × 2 = 5	1 × 2 = 2	0.25 × 4 = 1

Frequency = frequency data × class width

The number of cars in the sample = total frequency
= 3 + 5 + 5 + 2 + 1
= 16

Worked example

The frequency distribution below shows the height of 50 bushes, measured to the nearest centimetre.

Height (cm)	Frequency
10–14	3
15–19	6
20–24	7
25–29	9
30–39	12
40 49	8
50–74	4
75–99	1

Draw a histogram to represent this information.

First draw up a table to calculate the respective frequency densities, and then draw the histogram.

Since the height is measured to the nearest centimetre, the 10–14 interval extends from 9.5 to 14.5, giving it a width of 5 cm, and so on.

Height (cm)	Frequency	Class width	Frequency density
10–14	3	5	3 ÷ 5 = 0.6
15–19	6	5	6 ÷ 5 = 1.2
20–24	7	5	7 ÷ 5 = 1.4
25–29	9	5	9 ÷ 5 = 1.8
30–39	12	10	12 ÷ 10 = 1.2
40–49	8	10	8 ÷ 10 = 0.8
50–74	4	25	4 ÷ 25 = 0.16
75–99	1	25	1 ÷ 25 = 0.04

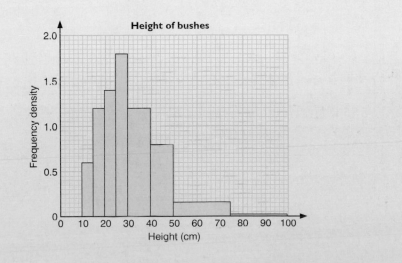

Height of bushes

≫ Frequency polygons

Frequency polygons were first discussed in Revision session 1, *Collecting and representing data*. A frequency polygon can be drawn from a histogram by joining the midpoints of the tops of the bars and extending the lines to the horizontal axis on either side. The area under the frequency polygon should be the same as the area under the histogram.

Worked example

The heights of people queuing for a fairground ride were as shown in the table.

Height (inches)	Frequency
35 up to 45	8
45 up to 55	13
55 up to 60	20
60 up to 65	29
65 up to 70	23
70 up to 90	11

Draw a frequency polygon to represent this information.

First draw up a table to calculate the respective frequency densities.

Height (inches)	Frequency	Class width	Frequency density
35 up to 45	8	10	0.8
45 up to 55	13	10	1.3
55 up to 60	20	5	4
60 up to 65	29	5	5.8
65 up to 70	23	5	4.6
70 up to 90	11	20	0.55

Then draw the histogram and frequency polygon.

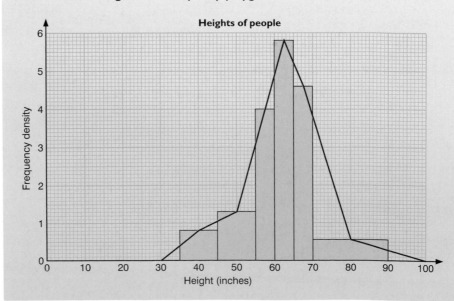

Q1 The following table shows the lengths of time taken by 120 workers to travel home one evening.

Time (t minutes)	Frequency
$0 \leqslant t < 10$	8
$10 \leqslant t < 20$	17
$20 \leqslant t < 30$	23
$30 \leqslant t < 60$	42
$60 \leqslant t < 90$	18
$90 \leqslant t < 120$	9
$120 \leqslant t < 240$	3

Draw a histogram to represent these data.

Q2 The following information shows the distance travelled by 100 salespeople one week. The distance is measured in miles to the nearest mile.

Distance (miles)	Frequency
0–500	3
501–1000	19
1001–2000	27
2001–4000	36
4001–6000	15

Use this information to construct:

a) a histogram

b) a frequency polygon.

Q3 The distance travelled by 50 lecturers to work is shown in the histogram below.

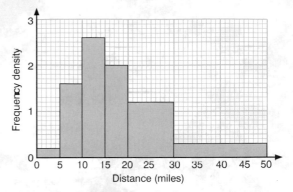

Use the information in the histogram to complete the following table.

Distance (miles)	0–	5–	10–	15–	20–	30–50
Number of lecturers						

Answers are on page 223.

Multiplication rule for dependent events

≫ What are dependent events?

Two or more events are **dependent** if one event affects the probability of the other event.

≫ The multiplication rule

Worked example

A bag contains four red and three blue counters. A counter is drawn from the bag and then a second counter is drawn from the bag. Draw a tree diagram to show the various possibilities that can occur and use the diagram to find the probability that both counters are blue.

The question does not make it clear whether the first counter is replaced before the second counter is drawn. This gives rise to two possibilities as shown in the following tree diagrams.

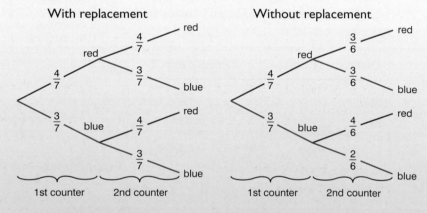

With replacement (independent events)

If the first counter is replaced before the second counter is drawn, then the two events are independent and the probabilities for each event are the same.

From the diagram, the probability that both counters are blue

= p(blue counter drawn first and blue counter drawn second)

= p(blue counter drawn first) × p(blue counter drawn second)

$= \frac{3}{7} \times \frac{3}{7} = \frac{9}{49}$

Without replacement (dependent events)

If the first counter is not replaced before the second counter is drawn, then the two events are not independent (i.e. they are dependent) and the probabilities on the second event will be affected by the outcomes on the first event.

For the second counter:

- if the first counter was blue, there are now two blue counters and six counters altogether

- if the first counter was not blue, there are still three blue counters and six counters altogether.

From the diagram you can see that:

the probability that both counters are blue

= p(blue counter drawn first and blue counter drawn second)

= p(blue counter drawn first) $\times$ p(blue counter drawn second)

$= \frac{3}{7} \times \frac{2}{6} = \frac{6}{42} = \frac{1}{7}$

CHECK YOURSELF QUESTIONS

Q1 Jack has ten black and six brown socks in his drawer. If he removes two socks from the drawer, one after the other, calculate the probability that:

a) both socks are black

b) both socks are brown

c) the socks are different colours.

Q2 The probability that Rebecca passes the driving theory test on her first attempt is $\frac{6}{7}$. If she fails then the probability that she passes on any future attempt is $\frac{7}{8}$.

Draw a tree diagram to represent this situation and use it to calculate the probability that Rebecca passes the driving test on her third attempt.

Answers are on page 224.

EXAM PRACTICE

Non-calculator paper

Questions 1–16 are intermediate tier (grades F to B). Questions 7–22 are higher tier (grade C to A*) . Try the questions for whichever tier you will be sitting in your actual exam. Answers and guidance are on pages 225–227.

1 Work out $5\frac{1}{5} - 1\frac{2}{3}$.

2 The population of the Earth is approximately 5×10^9 and the Earth's surface area is approximately $4 \times 10^{11} km^2$. Calculate the approximate area, in km^2, per head of population. Give your answer in standard form.

3 Draw accurately the net of this prism. Give your answer on a clean sheet of paper.

4 cm   4 cm   4 cm   7.5 cm

4 Give an approximate answer to: $\dfrac{28.65 \times 0.0852}{14.6 \times 3.22}$

by rounding the numbers to a sensible degree of accuracy.

5 The following data shows the length of paragraphs (in words) of two articles.

Article A   42 36 37 31 41 30 38 39 46 42 38 33 28 29 34 36 40 41
Article B   32 43 38 39 50 50 39 45 47 51 36 41 43 44 45 44 47 49

a) Show this information as a stem and leaf diagram.
Give your answer on a clean sheet of graph paper.
b) What do you notice about the two sets of data?

6 The equation of the straight line $y = mx + c$ is satisfied at the points $(2, 3)$ and $(1, {}^-2)$. What is the equation of the straight line?

7 A sequence of numbers is shown here.

Term	1st	2nd	3rd	4th	5th
Sequence	2	8	18	32	50

a) Write down the next term of the sequence

b) Write down the nth term of the sequence in terms of n.

c) Evaluate the 50th term of the sequence.

8 Factorise fully the expression $2\pi rh + 2\pi r^2$.

9 XYZ is an equilateral triangle of side 4.5 cm. Show all of the points which are less than 2.5 cm from the edges of the triangle. Give your answer on a clean sheet of paper.

10 The diameter of a £1 coin is 22 mm. The coin is 3 mm thick. Work out the volume of the coin, giving your answer in terms of π.

11 A rectangle measures 14 cm by 8 cm. A similar rectangle has sides 4 cm and x cm. What are the possible values of x?

12 Explain why the expression $\frac{4}{3}\pi r^2$ cannot represent the volume of a sphere.

13 This table shows the cumulative frequency for the test results of 72 students.

Age group (years)	10–	20–	30–	45–	50–	70–100
Frequency	2	0	6	4	22	12

From the table calculate:
a) how many students got marks less than or equal to 20

b) how many students got marks more than 40.

Draw the cumulative frequency curve for the data on a clean sheet of graph paper and use it to calculate:
c) the median and the interquartile range.

In the next test, also marked out of 50, the interquartile range was 30.
d) Comment on the two tests, using the interquartile ranges.

14 Solve these simultaneous equations.

$4x - y = 13$
$3x + y = 15$

15 Vijay and Baljit are playing a game with two fair five-sided spinners, one red and one blue. The blue spinner is numbered 5, 6, 7, 8, 9 and the red spinner is numbered 1, 2, 3, 4, 5. The final score is calculated by multiplying the two spinner scores together.

a) Complete a grid to show all the possible final scores.

b) Find the probability that the final score is a square number.

c) Find the probability that the final score is less than 30.

16 In the diagram, RT and PT are tangents to the circle. Calculate:

a) ∠ROP

b) ∠RSP

c) ∠RQP.

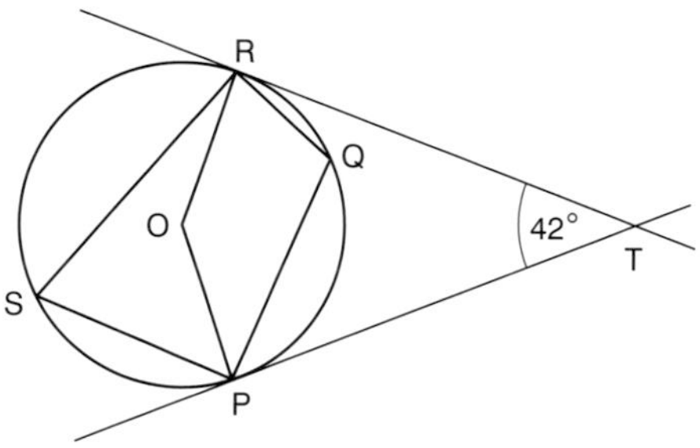

17 Simplify the following expressions, leaving your answers in surd form.

a) $\sqrt{12} + \sqrt{3}$

b) $\sqrt{12} \times \sqrt{6}$

18 Show that $0.2\dot{3}\dot{4}$ is a rational number.

19 The age of each person in a coach party is recorded.
The table shows the number of people in each age category.
Draw a histogram to represent the data.
Give your answer on a clean sheet of graph paper.

Age group (years)	10–	20–	30–	45–	50–	70–100
Frequency	2	0	6	4	22	12

20 A formula used by scientists is $t = \dfrac{v + p}{v}$.

Change the subject of the formula to v.

21 Write as a single fraction $\dfrac{1}{x + 3} + \dfrac{1}{x - 4}$.

22 The graph of $y = f(x)$ where $f(x) = \dfrac{x}{x + 1}$ is sketched below.

a) Sketch $y = f(x - 1)$.

b) Sketch $y = f(2x)$.

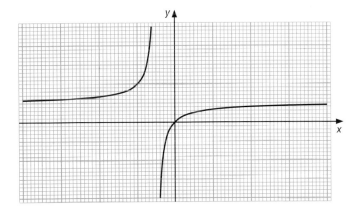

Questions 1–16 are intermediate tier (grades F to B). Questions 7–22 are higher tier (grade C to A*). Try the questions for whichever tier you will be sitting in your actual exam. Answers and guidance are on pages 228–230.

1 The speed of traffic on a three-lane stretch of road is in the ratio 2 : 3 : 5. If the speed of the traffic in the fastest lane is 60 miles per hour calculate the speed of traffic in the other two lanes.

2 Find the reciprocal of 0.25.

3 A questionnaire for schoolchildren includes the following question.

> 'How much pocket money do you get?'
>
> Less than £1 ☐   Between £2 and £5 ☐   More than £5 ☐

Write down two criticisms of this question.

4 The frequency distribution shows the different types of books borrowed from a library one weekend.
Show this information as a pie chart.

Type of book	Frequency
Sport	12
Crime	31
Horror	29
Romance	34
Other	14

5 The travel graph shows a train journey between two towns A and C stopping at B.

Use the graph to find:
a) the average speed between towns A and B

b) the average speed between towns A and C.

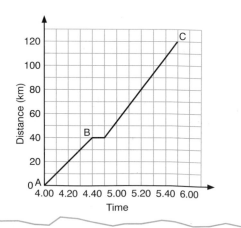

6 Solve the inequality $2(3 + 2x) > 4$ and show the solutions on a number line.

7 Use your calculator to work out the value of:

$$\frac{5.06 \times (10.32)^2}{281 + 217}$$

a) Write down the full calculator display.

b) Write down a calculation that could be carried out mentally to check this answer, using numbers rounded to one significant figure.

c) Write down the answer to your calculation in part (b).

8 A house is valued at £576 720 which represents an 8% increase on the original value. What was the original value?

9 A solution of the equation $x^3 + x = 12$ lies between 2 and 3. Use the method of trial and improvement to find this solution of the equation. Give your answer to 1 decimal place.

10 The formula for the surface area of a sphere, with area A cm^2 and radius r cm, is given as $A = 4\pi r^2$.

a) Find A when $r = 20$ cm.

b) Rearrange the formula to make r the subject.

c) Use this formula to find r when $A = 100$ cm^2.

11 Simplify the expression $(x + 5)^2 - (x - 5)^2$.

12 Two sides AB and DC of a regular pentagon ABCDE when produced meet at a point P. Calculate $\angle BPC$.

13 A washer has an outside diameter of 12 mm and an inside diameter of 6 mm. Calculate the cross-sectional area of the washer.

14 A vertical cliff is 485 metres high. The angle of depression of a boat at sea is 20°. What is the distance of the boat from the foot of the cliff?

15 ABC and ADE are straight lines. CE is a diameter.
Angle DCE = $x°$ and angle BCD = $2x°$.
Find, in terms of x, the sizes of the angles:

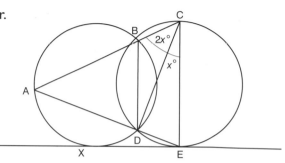

a) ABD

b) DBE

c) BAD.

16 Dr Malik wants to buy a car. She collects information about engine capacity and fuel consumption as shown in the table.

Engine capacity (litres)	Fuel consumption (mpg)
2.6	20.0
2.0	27.5
1.2	34.5
1.6	32.0
3.0	19.5
1.8	28.5
1.4	33.5

a) Plot this information and draw the line of best fit on the scatter graph.
Dr Malik decides to use her line to estimate the fuel consumption for other engine capacities.

b) Use the line of best fit to estimate the fuel consumption of a car with an engine capacity of:
i) 2.3 litres
ii) 3.5 litres.

c) Explain why one of these two estimates is more reliable than the other.

17 The value of a car, £v, is inversely proportional to its age, a years. After 1 year a car has a value of £7000. Find:

a) the value of the car after $3\frac{1}{2}$ years

b) the age of the car when it is worth £2500.

18 A bag contains four red and six blue balls. One ball is chosen and its colour noted. It is not put back into the bag. A second ball is chosen and its colour noted.

a)   Draw a tree diagram to represent this situation.

b) i)   Find the probability of obtaining two red balls.

    ii) Find the probability of obtaining one ball of each colour.

19 A windscreen wiper of length 25cm sweeps out an angle of 110° as illustrated in the diagram. What is the area of the screen covered?

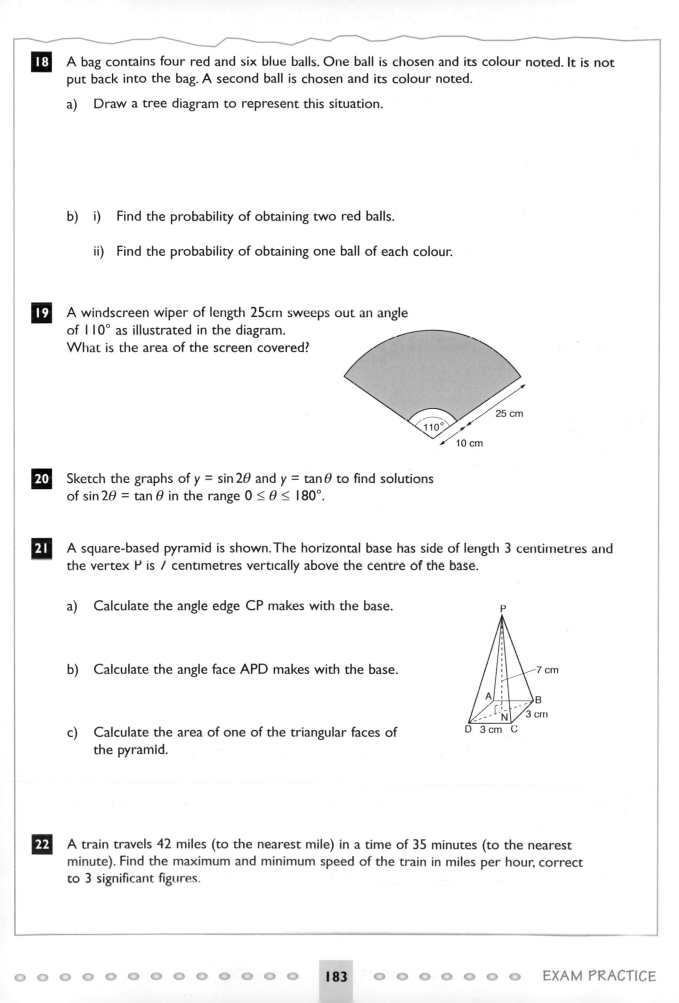

25 cm

110°

10 cm

20 Sketch the graphs of $y = \sin 2\theta$ and $y = \tan \theta$ to find solutions of $\sin 2\theta = \tan \theta$ in the range $0 \le \theta \le 180°$.

21 A square-based pyramid is shown. The horizontal base has side of length 3 centimetres and the vertex P is 7 centimetres vertically above the centre of the base.

a)   Calculate the angle edge CP makes with the base.

b)   Calculate the angle face APD makes with the base.

c)   Calculate the area of one of the triangular faces of the pyramid.

P

7 cm

A

B

N   3 cm

D   3 cm   C

22 A train travels 42 miles (to the nearest mile) in a time of 35 minutes (to the nearest minute). Find the maximum and minimum speed of the train in miles per hour, correct to 3 significant figures.

UNIT 1: NUMBER
1 The number system (page 2)

Q1
a) 7, 14 or 21
b) 1, 2, 5 or 10
c) 1, 3 or 17
d) 16 or 25
e) 17, 19 or 23
f) 2
g) 21

Comments
a) The numbers 7, 14 and 21 are all multiples of 7 and appear in the table.
b) The numbers 1, 2, 5 and 10 all divide exactly into 10 without a remainder.
c) The numbers 1, 3 and 17 all divide exactly into 51 and appear in the table.
d) The square numbers are 1, 4, 9, 16, 25, 36, 49, ... and the numbers 16 and 25 are both bigger than 10 and appear in the table.
e) The numbers 2, 3, 5, 7, 11, 13, 17, 19, 23, 29, 31, 37, ... are all prime numbers (i.e. they each only have two factors) although only 17, 19 and 23 are bigger than 16 and appear in the table.
f) The number 2 is the only prime number which is even. All other even numbers will have at least three factors i.e. 1, 2 and the number itself.
g) The numbers 3, 6, 9, 12, 15, 18, 21, 24, 27, ... are multiples of 3 and the numbers 7, 14, 21, 28, 35, 42, ... are multiples of 7. Of the numbers that appear in the table, only 21 is a multiple of 3 and also a multiple of 7.

Q2
$$264 = 2 \times 2 \times 2 \times 3 \times 11$$
$$\text{or} \quad 264 = 2^3 \times 3 \times 11$$

Comments
Use the factor tree method.

264
= 2×132 Writing 264 as the product 2×132
= $2 \times 2 \times 66$ Writing 132 as the product 2×66
= $2 \times 2 \times 6 \times 11$ Writing 66 as the product 2×33
= $2 \times 2 \times 2 \times 3 \times 11$
    Writing 6 as the product 2×3

2 Directed numbers (page 3)

Q1
a) $^+6$ or 6
b) $^-6$
c) $^-7$
d) $^+5$ or 5
e) $^+1$ or 1
f) $^-13$
g) $^-6$

Comments
a) $^-2 + 8 = 6$ Start at $^-2$ and go up 8.
b) $^-9 + 3 = ^-6$ Start at $^-9$ and go up 3.
c) $^-2 - 5 = ^-7$ Start at $^-2$ and go down 5.
d) $^+7 - ^+2 = ^+7 - 2 = 5$ Remember, $- + = -$ so rewrite as $^+7 - 2$.
e) $^-3 - ^-4 = ^-3 + 4 = 1$ Remember, $- - = +$ so rewrite as $^-3 + 4$.
f) $^-11 + ^-2 = ^-11 - 2 = ^-13$ $+ - = -$
g) $^-10 - ^-4 = ^-10 + 4 = ^-6$ $- - = +$

Q2
a) $^-6$
b) $^-21$
c) $^+30$ or 30
d) $^+4$ or 4
e) $^-8$
f) $^-2.5$ or $^-2\frac{1}{2}$
g) $^+12$ or 12

Comments
a) $^-3 \times ^+2 = ^-6$ $- \times + = -$
b) $^+7 \times ^-3 = ^-21$ $+ \times - = -$
c) $^-6 \times ^-5 = ^+30$ or 30 $- \times - = +$
d) $^+12 \div ^+3 = ^+4$ or 4 $+ \div + = +$
e) $^+16 \div ^-2 = ^-8$ $+ \div - = -$
f) $^-10 \div ^+4 = ^-2.5$ $10 \div 4 = 2.5$ and the sign is $- \div + = -$
g) $^-4 \times ^+3 \times ^-1 = 12$ $^-4 \times ^+3 = ^-12$ and $^-12 \times ^-1 = ^+12$ or 12

3 Rounding – significant figures and decimal places (page 5)

Q1
$174.9 = 175$ (3 s.f.)
$174.9 = 170$ (2 s.f.)
$174.9 = 200$ (1 s.f.)

Comments
$174.9 = 175$ (3 s.f.)
174 are the first 3 s.f. and the next most significant figure is 9. As 9 is bigger than 5, add 1 to the previous digit, giving 5.
The number $175 = 180$ (2 s.f.) but you must round the original 174.9.

$174.9 = 170$ (2 s.f.)
17 are the first 2 s.f. and the next most significant figure is 4. As 4 is less than 5, leave the previous digit alone. Fill with 0s to keep the number at its correct size.

$174.9 = 200$ (1 s.f.)
1 is the first s.f. and the next most significant figure is 7. As 7 is bigger than 5, add 1 to the previous digit, giving 2. Fill with 0s to keep the number at its correct size.

Q2 699.06 = 699 (3 s.f.)
     699.06 = 700 (2 s.f.)
     699.06 = 700 (1 s.f.)

Comments

699.06 = 699 (3 s.f.)
699 are the first 3 s.f. and the next most
significant figure is 0.
As 0 is less than 5, leave the previous digit alone.

699.06 = 700 (2 s.f.)
Here the second 0 is being included as a
significant figure.
69 are the first 2 s.f. and the next most significant
figure is 9.
As 9 is bigger than 5, add 1 to the previous digit,
giving 70.
Fill with 0s to keep the number at its correct size.

699.06 = 700 (1 s.f.)
6 is the first s.f. and the next most significant
figure is 9.
As 9 is bigger than 5, add 1 to the previous digit,
giving 7.
Fill with 0s to keep the number at its correct size.

Q3 0.8006 = 0.801 (3 d.p.)
     0.8006 = 0.80 (2 d.p.)
     0.8006 = 0.8 (1 d.p.)

Comments

0.8006 = 0.801 (3 d.p.)
800 are in the first 3 d.p. and the next digit is 6.
As 6 is bigger than 5, add 1 to the previous digit,
giving 801.
Rewrite the number, complete with decimal point,
replacing the 0 in the third d.p. by 1 and omitting
all the digits after it.

0.8006 = 0.80 (2 d.p.)
Here the second 0 is being included as a 'decimal
place holder'.
80 are in the first 2 d.p. and the next digit is 0.
As 0 is less than 5, leave the previous digit alone.
Rewrite the number, complete with decimal point,
omitting all the digits after the second decimal place.

0.8006 = 0.8 (1 d.p.)
8 is in the first d.p. and the next digit is 0.
As 0 is less than 5, leave the previous digit alone.
Rewrite the number, complete with decimal point,
omitting all the digits after the first decimal place.

4 Powers, roots and reciprocals (page 7)

Q1 $\sqrt{36} = \pm 6$, $\sqrt{10} = \pm 3.162\ 277\ 7$

Comments
The square root of a number is the number which,
when squared, gives that number.
$\sqrt{36} = \pm 6$ as $6 \times 6 = 36$, and $^-6 \times {}^-6 = 36$
(You should be able to do this without a calculator.)
$\sqrt{10}$ is not an exact number but lies between 3 and 4
(as $3 \times 3 = 9$ and $4 \times 4 = 16$). Using the $\boxed{\sqrt{}}$
key on your calculator gives $\sqrt{10} = 3.162\ 277\ 7$. You
must remember the negative square root as well.

Q2 $\sqrt[3]{4096} = 16$, $\sqrt[3]{-10} = {}^-2.1544347$

Comments
The cube root of a number is the number which,
when cubed, gives that number.
$\sqrt[3]{4096} = 16$ as $16 \times 16 \times 16 = 4096$.

$\sqrt[3]{10}$ is not an exact number but lies between
$^-2$ and $^-3$ (as $^-2 \times {}^-2 \times 2 = {}^-8$ and
$^-3 \times {}^-3 \times {}^-3 = {}^-27$).
Using the $\boxed{\sqrt[3]{}}$ key on your calculator gives
$\sqrt[3]{-10} + {}^-2.154\ 434\ 7$.

Q3 $\frac{4}{3}$, $\frac{1}{15}$, $\frac{5}{6}$

Comments
The reciprocal of $\frac{3}{4}$ is found by turning the fraction
upside-down to give $\frac{4}{3}$. The number 15 can be
written as $\frac{15}{1}$ and the reciprocal of $\frac{15}{1}$ is $\frac{1}{15}$.

Similarly the mixed number $1\frac{1}{5}$ can be written as $\frac{6}{5}$ (as
a top-heavy or an improper fraction – see Revision
session 7, Fractions) and the reciprocal of $\frac{6}{5}$ is $\frac{5}{6}$.

5 Positive, negative and zero indices (page 9)

Q1 a) 729   **b)** $\frac{1}{16}$ **c)** 6

Comments
a) An answer of 27 is a common mistake, caused
   by multiplying 9×3.
   9^3 tells you that three lots of 9 (the base
   number) are to be multiplied together
   (3 is the power or index).
   So $9^3 = 9 \times 9 \times 9 = 729$
b) 4^{-2} has a negative power.
   So $4^{-2} = \frac{1}{4^2} = \frac{1}{16}$ as $4^2 = 4 \times 4 = 16$
c) $6^1 = 6$ as any number raised to the power 1
   is always equal to that number.

Q2 a) 3^{23} **b)** 8^2 **c)** 1   **d)** 1600

Comments
a) $3^{11} \times 3^{12} = 3^{11 + 12} = 3^{23}$
b) $8^6 \div 8^4 = 8^{6 - 4} = 8^2$
c) $13^4 \div 13^4 = 13^{4 - 4} = 13^0 = 1$
d) As the base numbers are not the same you cannot use the rules of indices on this question and must work out the answer by working out 4^3 and 5^2.
$4^3 = 4 \times 4 \times 4 = 64$
$5^2 = 5 \times 5 = 25$
So $4^3 \times 5^2 = 64 \times 25 = 1600$

6 Standard form involving positive and negative indices (page 11)

Q1 a) 2.5×10^5 miles
   b) 1.67×10^{-21} milligrams

Comments
a) $A = 2.5$ and $n = 5$.
b) $A = 1.67$ and $n = {}^-21$. The negative value of n tells you that the number is very small.

Q2 a) 1.19×10^5 **b)** 7.81×10^{-4}
    c) 2.4×10^2 **d)** 5×10^6

Comments
a) $2.69 \times 10^5 - 1.5 \times 10^5 = (2.69 - 1.5) \times 10^5$
                     $= 1.19 \times 10^5$
b) $4.31 \times 10^{-4} + 3.5 \times 10^{-4} = 7.81 \times 10^{-4}$
c) $(6 \times 10^3) \times (4 \times 10^{-2})$
     $= (6 \times 4) \times 10^3 \times 10^{-2}$
     $= 24 \times 10^{3 - 2} = 24 \times 10^1$
     $= 2.4 \times 10^2$
     Rewriting in standard form where $24 = 2.4 \times 10^1$.
d) $(3 \times 10^{11}) \div (6 \times 10^4)$
     $= (3 \div 6) \times (10^{11} \div 10^4)$
     $= 0.5 \times 10^{11 - 4} = 0.5 \times 10^7$
     $= 5 \times 10^6$
     Rewriting in standard form where $0.5 = 5 \times 10^{-1}$.

Q3 9.45×10^{15} metres
    (to an appropriate degree of accuracy)

Comments
If light travels 2.998×10^8 metres in one second then it travels:
$2.998 \times 10^8 \times 60$ metres in one minute
$2.998 \times 10^8 \times 60 \times 60$ metres in one hour
$2.998 \times 10^8 \times 60 \times 60 \times 24$ metres in one day
$2.998 \times 10^8 \times 60 \times 60 \times 24 \times 365$ metres in one year.
$2.998 \times 10^8 \times 60 \times 60 \times 24 \times 365$
$= 9.454\,49 \times 10^{15}$ metres
$= 9.45 \times 10^{15}$ metres (to an appropriate degree of accuracy)

7 Fractions (page 15)

Q1 $2\frac{1}{20}$

Comments
$3\frac{1}{4} - 1\frac{1}{5}$
$= \frac{13}{4} - \frac{6}{5}$ Converting to top-heavy fractions.
$= \frac{65}{20} - \frac{24}{20}$ Writing both fractions with a denominator of 20.
$= \frac{41}{20} = 2\frac{1}{20}$ Rewriting as a mixed number..

Q2 $\frac{3}{10}$

Comments
$\frac{3}{\cancel{4}_2} \times \frac{\cancel{2}^1}{5} = \frac{3 \times 1}{2 \times 5}$ Cancelling fractions.
          $= \frac{3}{10}$

Q3 $4\frac{1}{2}$

Comments
$4\frac{4}{5} \div 1\frac{1}{15} = \frac{24}{5} \div \frac{16}{15}$ Converting to top-heavy fractions.
$= \frac{\cancel{24}^3}{\cancel{5}_1} \times \frac{\cancel{15}^3}{\cancel{16}_2}$ Multiplying by reciprocal and cancelling fractions.
$= \frac{9}{2}$
$= 4\frac{1}{2}$ Rewriting as a mixed number.

Q4 $\frac{81}{500}$

Comments
$0.162 = \frac{1}{10} + \frac{6}{100} + \frac{2}{1000}$
$= \frac{100}{1000} + \frac{60}{1000} + \frac{2}{1000}$
$= \frac{162}{1000}$
$= \frac{81}{500}$ Cancelling down.

8 Percentages (page 19)

Q1 23%

> **Comments**
>
> Fraction unemployed $= \dfrac{3 \text{ million}}{13 \text{ million}} = \dfrac{3}{13}$
>
> (as a fraction in its lowest terms).
> To convert the fraction to a percentage, multiply by 100 so the percentage $= \dfrac{3}{13} \times 100\% = 23.076\,923\%$
>
> $= 23\%$ (to the nearest whole number).

Q2 £2646

> **Comments**
>
> Percentage cost after discount is 94.5%.
>
> $100\% - 5.5\% = 94.5\%$
> $£2800 \times 94.5\% = £2800 \times 0.945$
> $ = £2646$

Q3 0.08%

> **Comments**
>
> Percentage error $= \dfrac{error}{actual\ amount} \times 100\%$
>
> Error $= 6.005 - 6 = 0.005$kg
>
> Percentage error $= \dfrac{0.005}{6.005} \times 100\% = 0.08\%$
>
> to an appropriate degree of accuracy.

9 Reverse percentages (page 20)

Q1 £380

> **Comments**
>
> After a reduction of 5%, the price represents 95% (100% − 5%) of the original cost of the holiday.
> 95% of the original cost of the holiday $= £361$
> 1% of the original cost of the holiday $= £\frac{361}{95} = £3.80$
> 100% of the original cost $= 100 \times £3.80 = £380$

Q2 £9500

> **Comments**
>
> After a depreciation of 45%, the price represents 55% (100% − 45%) of the original purchase price.
> 55% of original purchase price $= £5225$
> 1% of original purchase price $= £\frac{5225}{55} = £95$
> 100% of original price $= 100 \times £95 = £9500$

10 Estimation and approximation (page 21)

Q1 5

> **Comments**
>
> $\dfrac{3.87 \times 5.07^3}{5.16 \times 19.87} \approx \dfrac{4 \times 125}{5 \times 20} = \dfrac{500}{100} = 5$
>
> Take 5.07^3 as approximately $5^3 = 125$.

Q2 1

> **Comments**
>
> $\dfrac{2.78 + \pi}{\sqrt{5.95 \times 6.32}} \approx \dfrac{3 + 3}{\sqrt{6 \times 6}} = \dfrac{6}{6} = 1$ Take $\pi \approx 3$.

Q3 10

> **Comments**
>
> $\dfrac{59.96}{40.21 + 19.86} + \sqrt{80.652} \approx \dfrac{60}{40 + 20} + \sqrt{81} = 1 + 9 = 10$

11 Ratios and proportional division (page 23)

Q1 25 : 3

> **Comments**
>
> To find the ratio of 5 km to 600 m, you need to express both parts in the same units. 5 km = 5000 m so the ratio is 5000 : 600 = 25 : 3 in its simplest form.

Q2 1 : 2.5 or 1 : $2\frac{1}{2}$

> **Comments**
>
> 2 : 5 = 1 : 2.5   Dividing both sides by 2 to make an equivalent ratio in the form 1 : n.

Q3 4 : 3

> **Comments**
>
> $\frac{1}{3} : \frac{1}{4} = 4 : 3$ This is found by multiplying both sides by 12 to make an equivalent ratio.

Q4 £12, £18 and £42

> **Comments**
>
> For the ratio 2 : 3 : 7, the number of parts $= 2 + 3 + 7 = 12$.
> The value of each part $= £72 \div 12 = £6$.
> The children raise £12 (2 parts at £6), £18 (3 parts at £6) and £42 (7 parts at £6 each).
> **Check** £12 + £18 + £42 = £72 as required.

12 Mental methods (page 25)

Q1 11 832

Comments

$$
\begin{array}{r}
232 \\
\times \quad 51 \\
\hline
11600 \\
+ \quad 232 \\
\hline
11832 \\
\end{array}
$$

11600 Multiplying by 50.
+ 232 Multiplying by 1.
11832 Adding.

Q2 0.2149

Comments

Ignoring the decimal point and finding 614×35:

$$
\begin{array}{r}
614 \\
\times \quad 35 \\
\hline
18420 \\
+ \quad 3070 \\
\hline
21490 \\
\end{array}
$$

18420 Multiplying by 30.
+ 3070 Multiplying by 5.
21490 Adding.

Count the number of digits after the decimal point in each number.
0.0614 has 4 digits (4 d.p.)
3.5 has 1 digit (1 d.p.)
The answer needs $4 + 1 = 5$ decimal places.
Place the decimal point to give 5 d.p. in the answer.
$0.0614 \times 3.5 = 0.214\,90$ or 0.2149

Q3 2700

Comments

$459 \div 0.17$ can be written as

$\dfrac{459}{0.17} = \dfrac{45\,900}{17}$ or $45\,900 \div 17$

So $459 \div 0.17 = 2700$.

$$
\begin{array}{r}
2700 \\
17\overline{)45900} \\
34 \\
\hline
119 \\
119 \\
\hline
0 \\
\end{array}
$$

Q4 72

Comments

$0.0936 \div 0.0013$ can be written as

$\dfrac{0.0936}{0.0013} = \dfrac{936}{13}$ or $936 \div 13$

So $0.0936 \div 0.0013 = 72$.

$$
\begin{array}{r}
72 \\
13\overline{)936} \\
91 \\
\hline
26 \\
26 \\
\hline
0 \\
\end{array}
$$

13 Using a calculator (page 27)

Q1 26.01   **Q2** 3.456 607 1

Q3 0.031 25   **Q4** 500

Q5 0.0483   **Q6** $\frac{9}{14}$ or 0.642 857 1

Q7 $1\frac{2}{9}$ or 1.222 22

Comments

Refer to your user manual if you have any difficulties with this work.

14 Compound measures (page 29)

Q1 a) 16 m/s   **b)** 57.6 km/h

Comments

a) Speed $= \dfrac{\text{distance}}{\text{time}} = \dfrac{160}{10} = 16$ m/s

b) For speed in km/h, you need to express the distance in kilometres and the time in hours.

160 m = 0.16 km   as 1 km = 1000 m
10 seconds = $\frac{1}{6}$ minutes   as 60 seconds
                      = 1 minute
         = $\frac{1}{360}$ hour   as 60 minutes
                      = 1 hour

Speed $= \dfrac{\text{distance}}{\text{time}} = \dfrac{0.16}{\frac{1}{360}} = 57.6$ km/h

Q2 10.5 miles

Comments

As the speed is given in miles per hour, you need to express the time in hours.
Time = 45 minutes = $\frac{3}{4}$ hour   or you could use
                                0.75 hours
Distance = speed $\times$ time $= 14 \times \frac{3}{4} = 10.5$ miles

Q3 18.1 g (3 s.f.)

Comments

As the density is measured in g/cm^3, you need to express the mass in grams and the volume in cm^3.
Volume $= 1.2 \times 1.2 \times 1.2$ cm^3 as each side of the cube is 12 mm and 12 mm = 1.2 cm
         $= 1.728$ cm^3
Using mass = density $\times$ volume:
   mass $= 10.5 \times 1.728$
           $= 18.1$ g (3 s.f.)

15 Simple and compound interest (page 31)

Q1 2.5 years or 2 years 6 months

Comments

The amount $A = P + I$ so the interest
$I = £5875 - £5000 = £875$

$I = \dfrac{PRT}{100}$

$875 = \dfrac{5000 \times 7 \times T}{100}$ Substituting $I = £875$,
$P = £5000$ and $R = 7\%$.

$875 = 350 \times T$

$T = \dfrac{875}{350} = 2.5$

The time is 2.5 years or 2 years and 6 months
(12 months = 1 year).

Q2 5.25% p.a.

Comments

$I = \dfrac{PRT}{100}$

$52.50 = \dfrac{250 \times R \times 4}{100}$ Substituting $I = £52.50$,
$P = £250$ and
$T = 4$ years.

$52.50 = 10 \times R$

$R = \dfrac{52.50}{10} = 5.25\%$

The rate is 5.25% per annum.

Q3 Total amount = £2247.20 and
compound interest = £247.20

Comments

Using the formula $A = P \times \left(1 + \dfrac{R}{100}\right)^{T}$

$A = 2000\left(1 + \dfrac{6}{100}\right)^{2}$ Substituting $P = £2000$,
$R = 6\%$ and $T = 2$ years.

$A = £2247.20$ Remember to interpret
2247.2 on your calculator
as £2247.20.

$A = P + I$
$£2247.20 = £2000 + I$
$I = £247.20$
The total amount is £2247.20 and the compound
interest is £247.20.

16 Rational and irrational numbers (page 32)

Q1 $(\sqrt{3})^2$ and $\sqrt{6\frac{1}{4}}$ are rational numbers.

$(\sqrt{3})^2 = \frac{3}{1}$ and $\sqrt{6\frac{1}{4}} = \frac{5}{2}$

Comments

$(\sqrt{3})^2 = 3$ Since $(\sqrt{3})^2 = 3 = \frac{3}{1}$ in the form $\frac{p}{q}$.

$\sqrt{6\frac{1}{4}} = \frac{5}{2}$ Since $\sqrt{6\frac{1}{4}} = \sqrt{\frac{25}{4}} = \frac{\sqrt{25}}{\sqrt{4}} = \frac{5}{2}$.

Q2 a) $5\sqrt{3}$ **b)** $3\sqrt{5}$ **c)** $\dfrac{\sqrt{7}}{7}$

Comments

a) $\sqrt{5} \times \sqrt{15} = \sqrt{75} = \sqrt{25 \times 3} = \sqrt{25} \times \sqrt{3} = 5\sqrt{3}$
b) $\sqrt{5} + \sqrt{20} = \sqrt{5} + \sqrt{4 \times 5} = \sqrt{5} + 2\sqrt{5} = 3\sqrt{5}$
c) $\dfrac{1}{\sqrt{7}} = \dfrac{1}{\sqrt{7}} \times \dfrac{\sqrt{7}}{\sqrt{7}} = \dfrac{\sqrt{7}}{7}$

Q3 $(4 + \sqrt{3})(4 - \sqrt{3}) = 13$

Comments

$(4 + \sqrt{3})(4 - \sqrt{3}) = 4 \times 4 - 4\sqrt{3} + 4\sqrt{3} - \sqrt{3}\sqrt{3}$
$= 16 - 3$
$= 13$

17 Recurring decimals (page 33)

Q1 $\dfrac{253}{999}$

Comments

Notice that $1000 \times 0.\dot{2}5\dot{3} = 253.253\,253\ldots$
            Multiplying both sides by 1000.
and $1 \times 0.\dot{2}5\dot{3} = 0.253\,253\ldots$
Subtracting: $999 \times 0.\dot{2}5\dot{3} = 253$

and $0.\dot{2}5\dot{3} = \dfrac{253}{999}$ Dividing both sides by 999.

Q2 $\dfrac{827}{990}$

Comments

Notice that $100 \times 0.8\dot{3}\dot{5} = 83.5353535\ldots$
            Multiplying both sides by 1000.
and $1 \times 0.8\dot{3}\dot{5} = 0.8353535\ldots$
Subtracting: $99 \times 0.8\dot{3}\dot{5} = 82.7$
and $0.8\dot{3}\dot{5} = \dfrac{82.7}{99}$ Dividing both sides by 99.

$= \dfrac{827}{990}$ Writing as a proper fraction.

18 Fractional indices (page 34)

Q1 a) 6   b) 4   c) 10

> **Comments**
> a) $36^{\frac{1}{2}} = \sqrt{36} = 6$
> b) $64^{\frac{1}{3}} = \sqrt[3]{64} = 4$
> c) $10\,000^{\frac{1}{4}} = \sqrt[4]{10\,000} = 10$

Q2 a) 25   b) $\dfrac{1}{25}$

> **Comments**
> a) $125^{\frac{2}{3}} = (125^{\frac{1}{3}})^2 = (5)^2 = 25$
>
> b) $125^{-\frac{2}{3}} = \dfrac{1}{125^{\frac{2}{3}}}$
> $= \dfrac{1}{\left(125^{\frac{1}{3}}\right)^2}$
> $= \dfrac{1}{5^2}$
> $= \dfrac{1}{25}$

Q3 $n = -\frac{1}{2}$

> **Comments**
> $4^{\frac{1}{2}} = 2$ and $4^{-\frac{1}{2}} = \frac{1}{2}$ so $n = -\frac{1}{2}$

19 Direct and inverse proportion (page 35)

Q1 a) $T = 90$ b) $W = 4$

> **Comments**
> If T is proportional to the positive square root of W then $T \propto \sqrt{W}$ and $T = k\sqrt{W}$.
> Since $T = 36$ when $W = 16$ then $36 = k\sqrt{16}$.
> $36 = k \times 4$
> i.e. $k = 9$
> The equation is $T = 9\sqrt{W}$.
> a) When $W = 100$ then:
> $T = 9\sqrt{100}$
> $= 9 \times 10$
> $= 90$
> a) When $T = 18$ then $18 = 9\sqrt{W}$
> $\sqrt{W} = 2$
> $W = 4$ as $\sqrt{4} = 2$

Q2 a) $V = \frac{1}{9}$

b) $V = \dfrac{3}{1000}$ or 0.003 or 3×10^{-3}

> **Comments**
> If V varies inversely as the cube of Y then $V \propto \dfrac{1}{Y^3}$ and $V = \dfrac{k}{Y^3}$.
> Since $V = \frac{3}{8}$ when $Y = 2$ then:
> $\dfrac{3}{8} = \dfrac{k}{2^8}$
> $\dfrac{3}{8} = \dfrac{k}{8}$
> $k = 3$
> The equation is $V = \dfrac{3}{Y^3}$.
>
> a) When $Y = 3$ then:
> $V = \dfrac{3}{3^3} = \dfrac{3}{27}$
> $= \dfrac{1}{9}$ Cancelling down.
> b) When $Y = 10$ then:
> $V = \dfrac{3}{10^3} = \dfrac{3}{1000}$ or 0.003 or 3×10^{-3}

20 Upper and lower bounds (page 36)

Q1 26.5225 cm^2 minimum
27.5625 cm^2 maximum

> **Comments**
> Area of square = length × length
> minimum length = 5.15 cm
> area = 5.15 × 5.15 = 26.5225 cm^2
> maximum length = 5.25 cm
> area = 5.25 × 5.25 = 27.556 25 cm^2

Q2 £1300.01

> **Comments**
> As each price is given to the nearest £100 then:
> For the Ford Scorpio at £18 700:
> maximum = £18 749.99
> Note that £18 750 would be rounded up to £18 800.
> minimum = £18 650.00
> You must always take care when working with money or ages to identify lower and upper bounds.
> For the Vauxhall Calibra at £17 300:
> maximum = £17 349.99
> Note that £17 350 would be rounded up to £17 400.
> minimum = £17 250.00
> Least possible difference in price
> $= S_{min} - C_{max}$
> $= £18\,650 - £17\,349.99$
> $= £1300.01$

Unit 2: Algebra

1 Patterns and sequences (page 39)

Q1 a) $3, 5, 7, 9, 11, ...$ **b)** $^-5, ^-2, 1, 4, 7, ...$

    c) $^-2, ^-2, 0, 4, 10, ...$ **d)** $\frac{1}{2}, \frac{2}{3}, \frac{3}{4}, \frac{4}{5}, \frac{5}{6}, ...$

Comments

a) $2 \times 1 + 1 = 3, 2 \times 2 + 1 = 5,$
$2 \times 3 + 1 = 7$, etc.

b) $3 \times 1 - 8 = ^-5, 3 \times 2 - 8 = ^-2,$
$3 \times 3 - 8 = 1$, etc.

c) $1 \times 1 - 3 \times 1 = ^-2, 2 \times 2 - 3 \times 2 = ^-2,$
$3 \times 3 - 3 \times 3 = 0$, etc.

d) $\frac{1}{1+1} = \frac{1}{2}, \frac{2}{2+1} = \frac{2}{3}, \frac{3}{3+1} = \frac{3}{4}$, etc.

Q2 nth term $= 5n - 2$

Comments

As the differences are constant then a linear rule
can be applied to the terms to find the nth term.
Each term is the previous term $+ 5$ so try $5n$ first.

Term	1	2	3	4
Sequence	3	8	13	18
Try $5 \times$ (term number)	5	10	15	20
Sequence	3	8	13	18
Difference	$^-2$	$^-2$	$^-2$	$^-2$
The difference is always 2.				
So try $5 \times$ (term number) $- 2$.	3	8	13	18

So the nth term is $5 \times n - 2$ or $5n - 2$.

Q3 a) 10th term $= \frac{40}{11}$

    b) 100th term $= \frac{400}{101}$

    c) The value gets closer and closer to 4.

Comments

a) 10th term $= \frac{4 \times 10}{10 + 1}$

$\qquad = \frac{40}{11}$

b) 100th term $= \frac{4 \times 100}{100 + 1}$

$\qquad = \frac{400}{101}$

c) Trying out values of n as n increases in size,
you should notice that the value gets closer
and closer to 4.

2 Substitution (page 40)

Q1 33π cm^2

Comments

Substituting $r = 3$ (as the radius is 3 cm) and $l = 8$
(as the slant height is 8 cm) in the equation:
$A = \pi r l + \pi r^2$
$A = \pi \times 3 \times 8 + \pi \times 3 \times 3$ As $r^2 = r \times r$
$A = \pi \times 24 + \pi \times 9$

$\qquad$ Using the fact that
$\qquad \pi \times 24 + \pi \times 9 = \pi \times (24 + 9) = \pi \times 33$
$A = 33\pi$ cm^2

Q2 a) $u = 8$ **b)** $u = 7.5$ or $7\frac{1}{2}$

Comments

a) Substituting $f = 4$ and $v = 8$ in the formula
$\frac{1}{u} = \frac{1}{f} - \frac{1}{v}$:
$\frac{1}{u} = \frac{1}{4} - \frac{1}{8}$
$\frac{1}{u} = \frac{1}{8}$ As $\frac{1}{4} - \frac{1}{8} = \frac{1}{8}$
$u = 8$

b) Substituting $f = 3$ and $v = 5$ in the formula
$\frac{1}{u} = \frac{1}{f} - \frac{1}{v}$:
$\frac{1}{u} = \frac{1}{3} - \frac{1}{5}$
$\frac{1}{u} = \frac{2}{15}$ As $\frac{1}{3} - \frac{1}{5} = \frac{5}{15} - \frac{3}{15} = \frac{2}{15}$

$\qquad$ – see Number: Revision session 7,
$\qquad$ Fractions.
$u = \frac{15}{2}$ Reciprocating both sides.
$u = 7.5$ or $7\frac{1}{2}$

3 Simplifying expressions (page 42)

Q1 a) $a + 6b - 5c$ **b)** $7x + 7y - xy$
    c) $x^3 - 6x^2 + 3x - 2$ **d)** $24xyz$
    e) $10a^3$ **f)** $16a^4b^3c^2$ **g)** $2mn^2$

Comments

a) $3a + 6b - 2a - 5c = 3a - 2a + 6b - 5c$
    Collecting like terms together along with their signs.
$\qquad\qquad = a + 6b - 5c$

b) $5x + 7y - 3xy + 2x + 2yx$
$\qquad\qquad = 5x + 2x + 7y - 3xy + 2xy$ Collecting
$\qquad\qquad\qquad$ like terms and rewriting $2yx$ as $2xy$.
$\qquad\qquad = 7x + 7y - xy$

c) $x^3 + 3x^2 - 9x^2 - 4x + 7x$ 2
$\qquad\qquad = x^3 - 6x^2 + 3x - 2$ Collecting like terms.

d) $3x \times 4y \times 2z = 3 \times 4 \times 2 \times x \times y \times z$
$$= 24xyz$$

e) $5a \times 2a^2 = 5 \times 2 \times a \times a^2$
$$= 10 \times a \times a \times a \quad \text{As } a^2 = a \times a.$$
$$= 10a^3$$

f) $(4abc)^2 \times a^2b = 4abc \times 4abc \times a^2b$
$$= 16a^2b^2c^2 \times a^2b$$
$$\qquad\qquad \text{Multiplying } 4abc \times 4abc.$$
$$= 16a^4b^3c^2$$

g) $8mn^3 \div 4n = \dfrac{8 \times m \times n \times n \times n}{4 \times n}$
$$= \dfrac{{}^2\cancel{8} \times m \times n \times n \times \cancel{n}^{\,1}}{{}_1\cancel{4} \times \cancel{n}_1} = 2mn^2$$

4 Expanding and factorising (page 46)

Q1 a) $x - 3y$ **b)** $3a + 9b - 16c$
c) $2a^2 - ab + 2b^3$

Comments

a) $5x - (3y + 4x)$
$$= 5x - 1 \times 3y - 1 \times 4x$$
The $-$ outside the bracket is taken as $^-1$.
$$= 5x - 3y - 4x \quad \text{Multiplying out.}$$
$$= x - 3y \qquad \text{Collecting like terms together.}$$

b) $5(a + b - 2c) - 2(a - 2b + 3c)$
$$= 5 \times a + 5 \times b + 5 \times {}^-2c$$
$$\quad - 2 \times a - 2 \times {}^-2b - 2 \times 3c$$
$$= 5a + 5b - 10c - 2a + 4b - 6c$$
$$= 3a + 9b - 16c \quad \text{Collecting like terms together.}$$

c) $a(2a + b) - 2b(a - b^2)$
$$= a \times 2a + a \times b - 2b \times a - 2b \times {}^-b^2$$
$$= 2a^2 + ab - 2ab + 2b^3 \quad a \times a = a^2$$
$$\qquad\qquad\qquad ba \text{ is the same as } ab$$
$$\qquad\qquad\qquad {}^-2b \times {}^-b^2 = 2b^3$$
$$= 2a^2 - ab + 2b^3 \quad \text{Collecting like terms together.}$$

Q2 a) $2x(2x - 3)$ **b)** $2(lw + wh + hl)$
c) $5xy(x - 2y)$

Comments

a) $4x^2 - 6x = 2x(2x - 3)$
$2x$ is a common factor, $4x^2 = 2x \times 2x$ and $6x = 2x \times 3$.

b) $2lw + 2wh + 2hl = 2(lw + wh + hl)$
2 is the only factor common to all three terms.

c) $5x^2y - 10xy^2 = 5xy(x - 2y)$
$5xy$ is a common factor, $5x^2y = 5xy \times x$ and $10xy^2 = 5xy \times 2y$.

Q3 a) $x^2 + 5x + 4$ **b)** $6y^2 - 31y + 35$
c) $9x^2 + 6x + 1$

Comments

a) $(x + 1)(x + 4)$
$$= x \times x + x \times 4 + 1 \times x + 1 \times 4$$
$$= x^2 + 4x + 1x + 4$$
$$= x^2 + 5x + 4 \qquad \text{Collecting like terms.}$$

b) $(3y - 5)(2y - 7)$
$$= 3y \times 2y + 3y \times {}^-7 + {}^-5 \times 2y + {}^-5 \times {}^-7$$
$$= 6y^2 - 21y - 10y + 35 \quad \text{As } {}^-5 \times {}^-7 = {}^+35.$$
$$= 6y^2 - 31y + 35 \qquad \text{Collecting like terms.}$$

c) $(3x + 1)^2 = (3x + 1)(3x + 1)$
$$= 3x \times 3x + 3x \times 1 + 1 \times 3x + 1 \times 1$$
$$= 9x^2 + 3x + 3x + 1$$
$$= 9x^2 + 6x + 1 \qquad \text{Collecting like terms.}$$

Q4 a) $(x + 2)(x + 4)$ **b)** $(x + 1)(x - 2)$
c) $(x + 2)(x - 9)$

Comments

a) $x^2 + 6x + 8$
$= (x \quad)(x \quad)$ Look for numbers that multiply together to give a product of $^+8$.
      Try $^+1 \times {}^+8$ $^-1 \times {}^-8$
            $^+2 \times {}^+4$ $^-2 \times {}^-4$
$= (x + 2)(x + 4)$ Check by multiplying out.

b) $x^2 - x - 2$
$= (x \quad)(x \quad)$ Look for numbers that multiply together to give a product of $^-2$.
      Try $^-1 \times {}^+2$ $^+1 \times {}^-2$
$= (x + 1)(x - 2)$ Check by multiplying out.

c) $x^2 - 7x - 18$
$= (x \quad)(x \quad)$ Look for numbers that multiply together to give a product of $^-18$.
      Try $^-1 \times {}^+18$ $^+1 \times {}^-18$
            $^-2 \times {}^+9$ $^+2 \times {}^-9$
            $^-3 \times {}^+6$ $^+3 \times {}^-6$
$= (x + 2)(x - 9)$ Check by multiplying out.

5 Solving equations (page 48)

Q1 a) $x = 3.5$ **b)** $x = 1.5$ or $1\frac{1}{2}$
c) $z = 9$ **d)** $y = 5$
e) $x = 7$ **f)** $x = 5$
g) $y = 27$ **h)** $x = 7$

Comments

a) $3x = 10.5$ Subtracting 4.5 from both sides.

        $x = 3.5$ Dividing both sides by 3.

b) $x = \frac{30}{20} = 1.5$ or $1\frac{1}{2}$

c) $27 = 3z$ Multiplying both sides by z.

        $z = 9$ Dividing both sides by 3.

d) $3y = 20 - y$ Subtracting 5 from both sides.

        $4y = 20$ Adding y to both sides.

        $y = 5$ Dividing both sides by 4.

e) $8x - 12 = 44$ Expanding the brackets.

        $8x = 56$ Adding 12 to both sides.

        $x = 7$ Dividing both sides by 8.

f) $4x + 1 = 3x + 6$ Expanding the brackets.

        $4x = 3x + 5$ Subtracting 1 from both sides.

        $x = 5$ Subtracting 3x from both sides.

g) $y + 3 = 30$ Multiplying both sides by 10.

        $y = 27$ Subtracting 3 from both sides.

h) $6x - 12 - 4x - 2 = 0$

Remember the − sign outside the second bracket.

        $2x - 14 = 0$ Collecting like terms.

        $2x = 14$ Adding 14 to both sides.

        $x = 7$ Dividing both sides by 2.

Q2 25, 26 and 27

Comments

Let the three consecutive numbers be x, $x + 1$ and $x + 2$, then:

$x + (x + 1) + (x + 2) = 78$

     $3x + 3 = 78$ Collecting like terms.

     $3x = 75$ Subtracting 3 from both sides.

     $x = 25$ Dividing by 3.

So the numbers are 25, 26 and 27.

Q3 $x = 25$

Comments

The angles of a triangle add up to 180° so:

$x° + 2x° + (3x + 30)° = 180°$ and $x = 25$

6 Rearranging formulae (page 49)

Q1   **a)** $r = \dfrac{C}{2\pi}$ **b)** $u = v - at$

     c) $a = \dfrac{v - u}{t}$ **d)** $r = \sqrt{\dfrac{A}{\pi}}$

     e) $h = \dfrac{V}{\pi r^2}$ **f)** $r = \sqrt{\dfrac{V}{\pi h}}$

     g) $T = \dfrac{100I}{PR}$

Comments

a) $r = \dfrac{C}{2\pi}$ Dividing both sides by 2π

b) $u = v - at$ Subtracting at from both sides.

c) $at = v - u$ Subtracting u from both sides.

     $a = \dfrac{v - u}{t}$ Dividing both sides by t.

d) $r^2 = \dfrac{A}{\pi}$ Dividing both sides by π.

     $r = \sqrt{\dfrac{A}{\pi}}$ Taking the square root of both sides.

e) $h = \dfrac{V}{\pi r^2}$ Dividing both sides by πr^2.

f) $r^2 = \dfrac{V}{\pi h}$ Dividing both sides by πh.

     $r = \sqrt{\dfrac{V}{\pi h}}$ Taking the square root of both sides.

g) $100I = PRT$ Multiplying both sides by 100.

     $I = \dfrac{100I}{PR}$ Dividing both sides by PR.

Q2 $b = \sqrt{\dfrac{m}{a}}$

Comments

$a = \dfrac{m}{b^2}$

$ab^2 = m$ Multiplying both sides by b^2.

$b^2 = \dfrac{m}{a}$ Dividing both sides by a.

$b = \sqrt{\dfrac{m}{a}}$ Taking the square root of both sides.

7 Algebraic indices (page 50)

Q1   **a)** x^5 **b)** y^6

     c) $8a^{12}$ **d)** d^3

     e) $2x^3$ **f)** $\frac{1}{3}x^{-2}$ or $\dfrac{1}{3x^2}$

Comments

a) $x^4 \times x = x^{4+1} = x^5$ As $x = x^1$.

b) $(y^3)^2 = y^3 \times y^3 = y^{3+3} = y^6$

c) $(2a^4)^3 = (2a^4) \times (2a^4) \times (2a^4)$
$$= 2 \times a^4 \times 2 \times a^4 \times 2 \times a^4 \text{ Removing the}$$
$$\text{brackets.}$$
$$= 8 \times a^4 \times a^4 \times a^4 \qquad \text{As } 2 \times 2 \times 2 = 8.$$
$$= 8 \times a^{4+4+4} = 8a^{12}$$

d) $d^{12} \div d^9 = d^{12-9} = d^3$

e) $6x^7 \div 3x^4 = 2x^{7-4} = 2x^3$

f) $3x^6 \div 9x^8 = \frac{1}{3}x^{6-8} = \frac{1}{3}x^{-2} \qquad \text{As } \frac{3}{9} = \frac{1}{3}.$

$x^{-2} = \frac{1}{x^2}$ so $\frac{1}{3}x^{-2}$ can be written as $\frac{1}{3x^2}$.

Q2 a) $x = 4$ **b)** $y = 2$ **c)** $x = 1$

Comments

a) $3^x = 81$ $3^4 = 81$ so $x = 4$

b) $2^{3y} = 64$ $2^6 = 64$ so $3y = 6$ or $y = 2$

c) $5^{2x+1} = 125$ $5^3 = 125$ so $2x + 1 = 3$

$2x = 2$ or $x = 1$

8 Interpreting graphs (page 53)

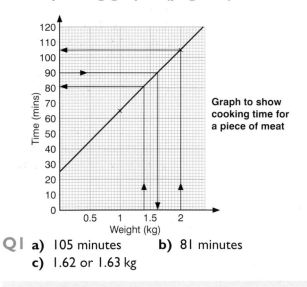

Graph to show cooking time for a piece of meat

Q1 a) 105 minutes   **b)** 81 minutes

    c) 1.62 or 1.63 kg

Comments

Reading from the graph:

a) A piece of meat weighing 2 kilograms would take 105 minutes to cook.

b) A piece of meat weighing 1.4 kilograms would take 81 minutes to cook.

c) A piece of meat which takes $1\frac{1}{2}$ hours to cook would weigh 1.62 or 1.63 kilograms. Converting $1\frac{1}{2}$ hours to 90 minutes.

Q2 a) 15 miles   **b)** 15 miles per hour

    c) 44 minutes   **d)** 1.19 p.m.

    e) 20 miles per hour   **f)** 2.03 p.m.

    g) 5.2 or 5.3 miles

Comments

Reading from the given graph:

a) The first cyclist travels 15 miles before the first stop.

b) Distance travelled = 15 miles

Time taken = 1 hour

So speed = 15 miles per hour

speed = distance ÷ time

c) The first cyclist stops for 44 minutes. Each small square represents 2 minutes.

d) The second cyclist overtakes the first cyclist at 1.19 p.m.

e) Distance travelled = 10 miles

Time taken = 30 minutes

So speed = distance ÷ time

$= 10 \div \frac{1}{2}$ As 30 minutes $= \frac{1}{2}$ hour.

$= 20$ miles per hour

f) The second cyclist arrives at the destination at 2.03 p.m.

g) The greatest distance between the two cyclists is 5.2 miles to 5.3 miles (when the second cyclist arrives at the destination).

9 Linear graphs and coordinates (page 55)

Q1

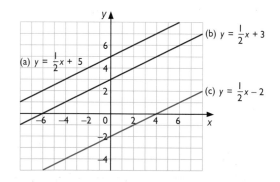

The lines are all parallel and have the same gradient.

Comments

Parallel lines have the same gradient and lines with the same gradient are parallel.

In this case the gradient, $m = \frac{1}{2}$.

Q2 a) $y + 3x = 5$ **b)** $x = 2y + 6$

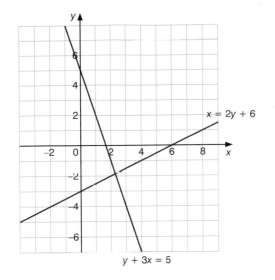

$x = 2y + 6$

$y + 3x = 5$

Comments

To sketch the graphs it is helpful to find the value of m (the gradient) and c (the cut-off on the y-axis, or y-intercept).

a) $y + 3x = 5$ Rearranging the formula to get it in the form $y = mx + c$.

$y = {}^-3x + 5$ Subtracting $3x$ from both sides.

$m = {}^-3$ and $c = 5$

b) $x = 2y + 6$ Rearranging the formula to get it in the form $y = mx + c$.

$x - 6 = 2y$ Subtracting 6 from both sides.

$2y = x - 6$ Turning the formula around.

$y = \frac{1}{2}(x - 6)$ Multiplying both sides by $\frac{1}{2}$ (or dividing both sides by 2).

$y = \frac{1}{2}x - 3$ Multiplying out the bracket.

$m = \frac{1}{2}$ and $c = {}^-3$.

Q3 a) $y = \frac{1}{2}x - 4$ **b)** $y = -\frac{5}{3}x - 5$

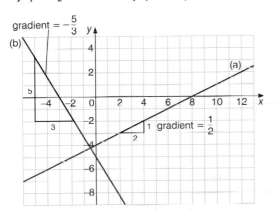

gradient $= -\frac{5}{3}$

(b)

(a)

gradient $= \frac{1}{2}$

Comments

a) From the graph you can see that $m = \frac{1}{2}$ and $c = {}^-4$.

so $y = mx + c$

$y = \frac{1}{2}x - 4$

b) From the graph you can see that $m = {}^-\frac{5}{3}$ and $c = {}^-5$.

so $y = mx + c$

$y = {}^-\frac{5}{3}x - 5$

10 Inequalities and graphs (page 58)

Q1 a $x \leqslant 3$

-3 -2 -1 0 1 2 3 4

b $x > 5$

-1 0 1 2 3 4 5 6 7

Comments

a) $4x + 2 \leqslant 17 - x$

$4x \leqslant 15 - x$ Subtracting 2 from both sides.

$5x \leqslant 15$ Adding x to both sides.

$x \leqslant 3$ Dividing both sides by 5.

b) $18 - 6x < 3 - 3x$

$18 \leqslant 3 + 3x$ Adding $6x$ to both sides.

$15 < 3x$ Subtracting 3 from both sides.

$5 < x$ Dividing both sides by 3.

$x > 5$ Rewriting the inequality to take x to the left-hand side.

Q2 a) $y \leqslant x$ **b)** $y < 7x + 2.5$

c) $y \leqslant {}^-x^2$ and $y \geqslant {}^-4$

Comments

a) $y \leqslant x$

Values can be taken either side of the given line to ascertain the required region.

b) $y < 7x + 2.5$

The dotted line shows that the line is **not** included in the required region.

c) $y \leqslant {}^-x^2$ and $y \geqslant {}^-4$

Both inequalities must be satisfied to give the required region.

Q3

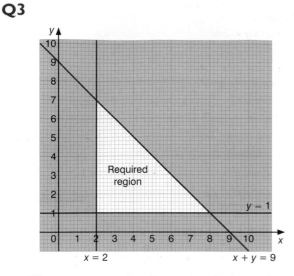

$x = 2$ $\qquad$ $x + y = 9$

The maximum value for $x + y$ which satisfies all of these conditions is 9 at the points $(2, 7)$ and $(8, 1)$ on the graph.

11 Quadratic graphs (page 61)

Q1

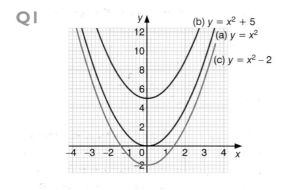

(a) $y = x^2$
(b) $y = x^2 + 5$
(c) $y = x^2 - 2$

Q2
a) Minimum $= (3, {}^-4)$
b) i) $x = 1$ and $x = 5$ $\qquad$ ii) $x = 0$ and $x = 6$

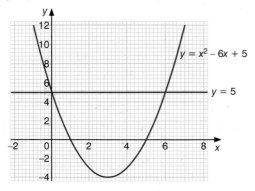

$y = x^2 - 6x + 5$
$y = 5$

Comments

From the graph you can see that:
a) the coordinates of the minimum value of $x^2 - 6x + 5$ are $(3, {}^-4)$

b) i) the values of x when $x^2 - 6x + 5 = 0$ will lie on $y = x^2 - 6x + 5$ and $y = 0$, so the values $x = 1$ and $x = 5$ satisfy the equation $x^2 - 6x + 5 = 0$
ii) the values of x when $x^2 - 6x + 5 = 5$ will lie on $y = x^2 - 6x + 5$ and $y = 5$, so the values $x = 0$ and $x = 6$ satisfy the equation $x^2 - 6x + 5 = 5$.

Q3

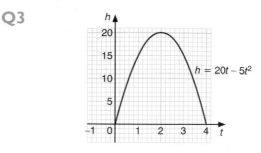

$h = 20t - 5t^2$

Maximum height = 20 metres

Comments

From the graph, the maximum height occurs at the point $(2, 20)$ so the maximum height is 20 metres (which occurs when the time is 2 seconds).

12 Cubic and reciprocal graphs (page 64)

Q1

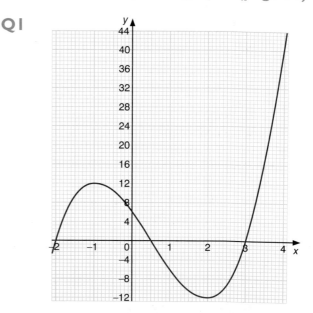

Comments

You should use a table of values in the given range to draw the graph, bearing in mind the general shape of a cubic curve.

x	${}^-2$	${}^-1$	0	1	2	3	4
y	0	12	6	${}^-6$	${}^-12$	0	42
Coordinates	$({}^-2, 0)$	$({}^-1, 12)$	$(0, 6)$	$(1, {}^-6)$	$(2, {}^-12)$	$(3, 0)$	$(4, 42)$

Q2

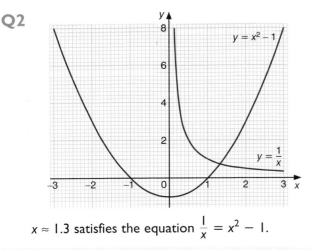

$x \approx 1.3$ satisfies the equation $\frac{1}{x} = x^2 - 1$.

Comments

The reciprocal curve, $y = \frac{1}{x}$ is not defined at $x = 0$. You will need to use non-integer values, between 0 and 1, to find out sufficient detail about the behaviour of the curve to be able to draw its graph. From the graph you can see that the curve $y = \frac{1}{x}$ crosses the curve $y = x^2 - 1$ when $x \approx 1.3$ so that $x \approx 1.3$ satisfies the equation $\frac{1}{x} = x^2 - 1$.

13 Simultaneous equations (page 66)

Q1 a) $x = 4$
and $y = 2$

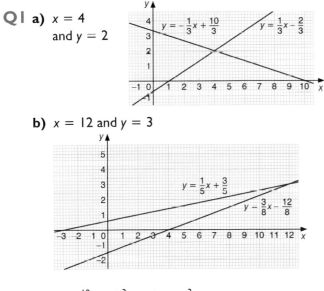

b) $x = 12$ and $y = 3$

c) $x = \frac{10}{7}$ or $1\frac{3}{7}$ and $y = \frac{3}{7}$

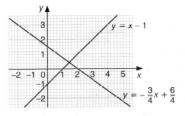

Comments

a) $x + 3y = 10$
$2x - 3y = 2$
 (i) Graphical method
  Plot $y = -\frac{1}{3}x + \frac{10}{3}$ and $y = \frac{2}{3}x - \frac{2}{3}$.
  Intersect at (4, 2).
 (ii) Substitution method
  Substitute $x = 10 - 3y$ in the second equation.
 (iii) Elimination method
  Add the two equations to eliminate the term in y.

b) $x = 5y - 3$
$3x - 8y = 12$
 (i) Graphical method
  Plot $y = \frac{1}{5}x + \frac{3}{5}$ and $y = \frac{3}{8}x - \frac{12}{8}$.
  Intersect at (12, 3).
 (ii) Substitution method
  Substitute $x = 5y - 3$ in the second equation.
 (iii) Elimination method
  Rewrite the first equation as $x - 5y = {}^-3$ and then $3x - 15y = {}^-9$ (multiplying by 3). Subtract the two equations to eliminate the term in x.

c) $x - 1 = y$
$3x + 4y = 6$
 (i) Graphical method
  Plot $y = x - 1$ and $y = \frac{{}^-3}{4}x + \frac{6}{4}$.
  Intersect at $(\frac{10}{7}, \frac{3}{7})$.
 (ii) Substitution method
  Substitute $y = x - 1$ in the second equation.
 (iii) Elimination method
  Rewrite the first equation as $x - y = 1$ and then $4x - 4y = 4$ (multiplying by 4). Add the two equations to eliminate the term in y.

Q2 16 and 20

Comments

Let the two numbers be x and y (say).
$x + y = 36$ As their sum is 36.
$x - y = 4$ As their difference is 4.
Solving these simultaneous equations by the elimination method and adding the two equations:
$(x + y) + (x - y) = 36 + 4$
$2x = 40$
$x = 20$
Substituting in the first equation:
$x + y = 36$
$20 + y = 36$ As $x = 20$.
$y = 16$
So the two numbers are 20 and 16.

Q3 The cost of a tie is £8.00 and the cost of a shirt is £16.50.

The cost of a tie and a shirt is £24.50.

Comments

Let £t be the cost of one tie and £s be the cost of one shirt.

$2t + s = 32.50$ As the cost of two ties and a shirt is £32.50.

$t + 2s = 41.00$ As the cost of one tie and two shirts is £41.00.

Solving these simultaneous equations by the substitution method:

Using $2t + s = 32.50$ you can write $s = 32.50 - 2t$.

Substituting this value of s into the second equation:

$t + 2s = 41$ 41 is the same as 41.00

$t + 2(32.50 - 2t) = 41$ Substituting $s = 32.50 - 2t$.

$t + 65 - 4t = 41$ Expanding the brackets.

$-3t = {}^-24$ Collecting like terms on each side.

$t = 8$ Dividing both sides by $^-3$.

You can now use $s = 32.50 - 2t$ with $t = 8$ to find s.

$s = 32.50 - 2t$

$s = 32.50 - 2 \times 8$ As $t = 8$.

$s = 16.50$

The cost of a tie is £8.00 and the cost of a shirt is £16.50.

A tie and a shirt together cost £24.50.

14 Quadratic equations (page 68)

Q1 a) $x = 5$ and $x = 7$ **b)** $x = 6$ and $x = -\frac{1}{2}$
c) $x = 1$ and $x = -5$

Comments

a) $(x - 5)(x - 7) = 0$ Either $(x - 5) = 0$
                              or $(x - 7) = 0$

The solutions are $x = 5$ and $x = 7$.

b) $(x - 6)(2x + 1) = 0$ Either $(x - 6) = 0$
                               or $(2x + 1) = 0$

The solutions are $x = 6$ and $x = -\frac{1}{2}$.

c) $x^2 + 4x + 5 = 0$

$(x - 1)(x + 5) = 0$ Either $(x - 1) = 0$
                        or $(x + 5) = 0$

The solutions are $x = 1$ and $x = {}^-5$.

Q2 a) $(x - 2)(x - 5) = 0$ or $x^2 - 7x + 10 = 0$
b) $(x + 3)(5x - 1) = 0$ or $5x^2 + 14x - 3 = 0$

Comments

Reverse the above process to find the factors.

a) $x = 2$ and $x = 5$

Either $(x - 2) = 0$ or $(x - 5) = 0$

$(x - 2)(x - 5) = 0$ or $x^2 - 7x + 10 = 0$

b) $x = {}^-3$ and $x = \frac{1}{5}$

Either $(x + 3) = 0$ or $(x - \frac{1}{5}) = 0$

$x - \frac{1}{5} = 0$ gives $5x - 1 = 0$ Multiplying both sides by 5.

$(x + 3)(5x - 1) = 0$ or $5x^2 + 14x - 3 = 0$

Q3 $x = 6$ and $x = {}^-1$

Comments

$x^2 - 5x + 2 = 8$

$x^2 - 5x - 6 = 0$ Writing in the form
                     $ax^2 + bx + c = 0$.

$(x - 6)(x + 1) = 0$ Factorising.

Either $(x - 6) = 0$ or $(x + 1) = 0$.

The solutions are $x = 6$ and $x = {}^-1$.

Q4 width $= 8$ metres, length $= 12$ metres

Comments

Let the width be x centimetres, then the length is $(x + 4)$ centimetres.

Area $= x(x + 4) = 96$

$x^2 + 4x = 96$ Expanding the brackets.

$x^2 + 4x - 96 = 0$ Writing in the form
                      $ax^2 + bx + c = 0$.

$(x - 8)(x + 12) = 0$ Factorising Either $x = 8$ or
                        $x = {}^-12$.

But a width of $^-12$ centimetres is impossible so ignore this value. Since the length is 4 cm greater than the width, this gives width $= 8$ cm and length $= 12$ cm.

15 Trial and improvement methods (page 70)

QI 4.6 (1 d.p.)

Comments

We know the solution lies between 4 and 5, so try 4.5 first.

When $x = 4.5$ $x^3 + x = 95.625$ (too small)

When $x = 4.6$ $x^3 + x = 101.936$ (too large)

When $x = 4.55$ $x^3 + x = 98.746\ 375$

So x lies between 4.55 and 4.6 which are both equivalent to 4.6 (1 d.p.).

Q2 1.9 (1 d.p.)

Comments

To find an approximate solution of the equation it is helpful to draw a graph of the curve $y = x^3 - x$ and the line $y = 5$ and note the solution where they cross. Alternatively, consider a few different values to find an appropriate starting point.

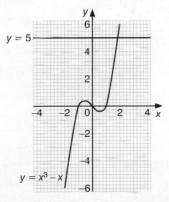

From the graph you can see that the solution lies between 1.8 and 2.

When $x = 1.8$ $x^3 - x = 4.032$ Substituting $x = 1.8$.

When $x = 2$ $x^3 - x = 6$ Substituting $x = 2$.

When $x = 1.9$ $x^3 - x = 4.959$ So x is between 1.9 and 2.

When $x = 1.95$ $x^3 - x = 5.464\ 875$ So x is between 1.9 and 1.95.

When $x = 1.92$ $x^3 - x = 5.157\ 888$

So x lies between 1.9 and 1.92 which are both equal to 1.9 (1 d.p.).

16 Further algebra (page 71)

QI $x = \frac{1}{4}$

Comments

$$1 + \frac{3x^2}{x - 1} = 3x \qquad \text{Multiply throughout by } x.$$

$$1(x - 1) + 3x^2 = 3x(x - 1) \quad \text{Multiply throughout by } (x - 1).$$

$$x - 1 + 3x^2 = 3x^2 - 3x \quad \text{Expanding.}$$

$$4x - 1 = 0 \qquad \text{Simplifying.}$$

$$x = \frac{1}{4}$$

Q2 $s = \dfrac{r^2 t}{1 - r^2}$

Comments

$$r^2 = \frac{s}{s + t} \qquad \text{Squaring both sides.}$$

$$r^2(s + t) = s \qquad \text{Multiplying both sides by } (s + t).$$

$$r^2 s + r^2 t = s \qquad \text{Expanding the brackets.}$$

$$s - r^2 s = r^2 t \qquad \text{Collecting together terms in } s \text{ on one side.}$$

$$s(1 - r^2) = r^2 t \qquad \text{Factorising the terms in } s.$$

$$s = \frac{r^2 t}{1 - r^2} \qquad \text{Dividing both sides by } (1 - r^2) \text{ to make } s \text{ the subject.}$$

Q3 a) $x = \frac{3}{5}$ or 0.6   b) $t = \sqrt{\dfrac{1 - x}{1 + x}}$

Comments

a)

$$x = \frac{1 - t^2}{1 + t^2}$$

$$x = \frac{1 - \left(\frac{1}{2}\right)^2}{1 + \left(\frac{1}{2}\right)^2} \quad \text{Substituting } \frac{1}{2} \text{ for } t.$$

$$x = \frac{\frac{3}{4}}{\frac{5}{4}} = \frac{3}{5}$$

b)

$$x = \frac{1 - t^2}{1 + t^2}$$

$$x(1 + t^2) = 1 - t^2 \quad \text{Multiplying both sides by } (1 + t^2).$$

$$x + xt^2 = 1 - t^2 \quad \text{Expanding the brackets.}$$

$$xt^2 + t^2 = 1 - x \quad \text{Collecting together terms in } t \text{ on one side only.}$$

$$t^2(x + 1) = 1 - x \quad \text{Factorising the terms in } t^2.$$

$$t^2 = \frac{1 - x}{1 + x} \quad \text{Dividing both sides by } x + 1 \text{ to make } t^2 \text{ the subject.}$$

$$t = \sqrt{\frac{1 - x}{1 + x}} \quad \text{Taking the square root of both sides to make } t \text{ the subject.}$$

17 Further simultaneous equations (page 73)

Q1 a) $x = 0.2, y = 0.2$ and $x = 2, y = 20$

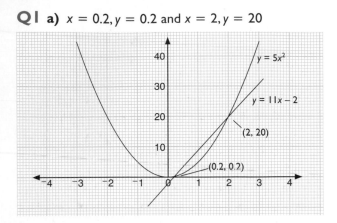

b) $x = 3, y = 4$ and $x = 4, y = 3$

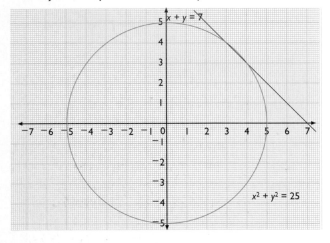

Comments

a) You can easily sketch the graphs of the two simultaneous equations.

The coordinates of the point of intersection, $(0.2, 0.2)$ and $(2, 20)$ give the solution of the simultaneous equations.

b) You can easily sketch the graphs of the two simultaneous equations.

The coordinates of the point of intersection, $(3, 4)$ and $(4, 3)$ give the solution of the simultaneous equations.

Q2 a) $x = 0.2, y = 0.2$ and $x = 2, y = 20$
   b) $x = 3, y = 4$ and $x = 4, y = 3$

Comments

a) To solve the simultaneous equations, substitute the value of $y = 11x - 2$ into $y = 5x^2$.

$11x - 2 = 5x^2$ Substituting $y = 11x - 2$.

$5x^2 - 11x + 2 = 0$ Collecting terms on one side.

$(5x - 1)(x - 2) = 0$ Factorising the quadratic equation.

Either $x = \frac{1}{5}$ or $x = 2$.

If $x = \frac{1}{5}$ then $y = \frac{1}{5}$.   As $y = 5x^2$.

If $x = 2$ then $y = 20$ As $y = 5x^2$ again.

b) To solve the simultaneous equations, use $x + y = 7$ to substitute the value $y = 7 - x$ into $x^2 + y^2 = 25$.

$x^2 + (7 - x)^2 = 25$ Substituting $y = 7 - x$.

$x^2 + (49 - 14x + x^2) = 25$ Expanding the brackets.

$x^2 + 49 - 14x + x^2 = 25$ Expanding the brackets.

$2x^2 - 14x + 24 = 0$ Collecting terms on one side.

$x^2 - 7x + 12 = 0$ Dividing throughout by 2.

$(x - 3)(x - 4) = 0$ Factorising the quadratic.

Either $x = 3$ or $x = 4$.

If $x = 3$ then $y = 4$.   As $y = 7 - x$.

If $x = 4$ then $y = 3$.   As $y = 7 - x$ again.

18 Further quadratic equations (page 76)

Q1 $x = -\frac{1}{2}$ and $x = {}^-2$

Comments

Comparing the given quadratic with the general form $ax^2 + bx + c = 0$ then: $a = 2, b = 5, c = 2$.

The equation will need to be rewritten to make the coefficient of x^2 equal to 1.

The quadratic can be rewritten as $x^2 + \frac{5}{2}x + 1 = 0$ (dividing throughout by 2)

Comparing this quadratic with the general form $ax^2 + bx + c = 0$ then: $a = 1, b = \frac{5}{2}, c = 1$

Since $a = 1$ then the 'square' is $(x + \frac{b}{2})^2$

The square is $(x + \frac{5/2}{2})^2 = (x + \frac{5}{4})^2$.

Now $(x + \frac{5}{4})^2 = x^2 + \frac{5}{2}x + \frac{25}{16}$

So $x^2 + \frac{5}{2}x = (x + \frac{5}{4})^2 - \frac{25}{16}$ Subtracting $\frac{25}{16}$ from both sides.

So $x^2 + \frac{5}{2}x + 1 = 0$ Using the original quadratic.

$(x + \frac{5}{4})^2 - \frac{25}{16} + 1 = 0$ Replacing $x^2 + \frac{5}{2}x$ by $(x + \frac{5}{4})^2 - \frac{25}{16}$.

$(x + \frac{5}{4})^2 - \frac{9}{16} = 0$ Simplifying.

$(x + \frac{5}{4})^2 = \frac{9}{16}$ Isolating the square term on the LHS.

$x + \frac{5}{4} = \sqrt{\frac{9}{16}}$ Taking square roots on both sides.

$x + \frac{5}{4} = \pm\frac{3}{4}$ Remember that the square root has two solutions.

If $x + \frac{5}{4} = \frac{3}{4}$ then $x = -\frac{1}{2}$.

If $x + \frac{5}{4} = -\frac{3}{4}$ then $x = -2$.

Q2 a) $x = 0.414$ or $x = -2.414$ (3 s.f.)
    b) $x = 4.56$ or $x = 0.438$ (3 s.f.)

Comments

a) Comparing $x^2 + 2x - 1 = 0$ with the general form $ax^2 + bx + c = 0$:

$a = 1, b = 2, c = -1.$

$x = \dfrac{-2 \pm \sqrt{2^2 - 4 \times 1 \times -1}}{2 \times 1}$

$x = \dfrac{-2 \pm \sqrt{8}}{2}$

$x = \dfrac{-2 + 2.828\,427\,125}{2}$

or $x = \dfrac{-2 - 2.828\,427\,125}{2}$

$x = 0.414$ or $x = -2.414$ (3 s.f.)

b) The equation $2x^2 = 10x - 4$ needs to be rearranged in order to compare it with the general form $ax^2 + bx + c = 0$.

$2x^2 = 10x - 4$

$2x^2 - 10x + 4 = 0$

Comparing $2x^2 - 10x + 4 = 0$ with the general form $ax^2 + bx + c = 0$:

$a = 2, b = -10, c = 4.$

$x = \dfrac{-10 \pm \sqrt{(-10)^2 - 4 \times 2 \times 4}}{2 \times 2}$

$x = \dfrac{10 \pm \sqrt{68}}{4}$ Remember that $- \times - = +$ and $(-10)^2 = 100.$

$x = \dfrac{10 + 8.246\,211\,251}{4}$

or $x = \dfrac{10 - 8.246\,211\,251}{4}$

$x = 4.56$ or $x = 0.438$ (3 s.f.)

Q3 width = 4 metres, length = 8 metres

Comments

Let the width be x metres, then the length is $(x + 4)$ metres.

Area $= x(x + 4) = 32$

$x^2 + 4x = 32$

$x^2 + 4x - 32 = 0$

Comparing $x^2 + 4x - 32 = 0$ with the general form $ax^2 + bx + c = 0$:

$a = 1, b = 4, c = -32.$

$x = \dfrac{-4 \pm \sqrt{4^2 - 4 \times 1 \times -32}}{2 \times 1} = \dfrac{-4 \pm \sqrt{144}}{2}$

$x = \dfrac{-4 + 12}{2} = 4$ or $x = \dfrac{-4 - 12}{2} = -8$

But a width of -8 metres is impossible so ignore this value. Since the length of the room is 4 m greater than the width: width = 4 m, length = 8 m. The fact that the square root worked out to be an exact number ($\sqrt{144} = 12$) suggests that the quadratic equation $x^2 + 4x - 32 = 0$ could have been factorised which would have made the problem a lot easier and quicker. As a rule, you should always use factorising rather than the formula, if you can.

Q4 $x = 10.0990$ (4 d.p.)

Comments

Using $x_{n+1} = 10 + \dfrac{1}{x_n}$ and $x_1 = 5$:

$x_2 = 10 + \dfrac{1}{x_1} = 10 + \dfrac{1}{5} = 10.2$

$x_3 = 10 + \dfrac{1}{x_2} = 10 + \dfrac{1}{10.2} = 10.098\,039\,22$

$x_4 = 10 + \dfrac{1}{x_3} = 10 + \dfrac{1}{10.098\,039\,22} = 10.099\,029\,13$

$x_5 = 10 + \dfrac{1}{x_4} = 10 + \dfrac{1}{10.099\,029\,13} = 10.099\,019\,42$

Since $x_5 = x_4$ to 4 d.p. then a root of the equation is 10.0990 (4 d.p.).

Q5 a) $x_2 = 0.75$ $x_3 = 1.043\,478\,3$ $x_4 = 0.992\,805\,7$
    b) The value of x_n would seem to be tending towards an answer of 1.
    c) $x = \dfrac{6}{x + 5}$

$x(x + 5) = 6$

$x^2 + 5x = 6$

$x^2 + 5x - 6 = 0$

    d) $x = 1$ or $x = -6$

Comments

a) Using $x_{n+1} = \dfrac{6}{x_n + 5}$ and $x_1 = 3$:

$x_2 = 0.75$ $x_3 = 1.043\,478\,3$ $x_4 = 0.992\,805\,7$

b) The value of x_n would seem to be tending towards an answer of 1 as n becomes very large.

c) The quadratic equation is found by rearranging:

$$x = \frac{6}{x + 5}$$

$x(x + 5) = 6$ Multiplying both sides
by $(x + 5)$.

$x_2 + 5x - 6 = 0$ Expanding and rearranging.

d) Factorising the left-hand side of the equation:

$x^2 + 5x - 6 = (x - 1)(x + 6)$

giving $(x - 1)(x + 6) = 0$.

So the solutions of the quadratic equation
$x^2 + 5x - 6 = 0$ are $x = 1$ and $x = {}^-6$.

19 Gradients and tangents (page 79)

Q1 $y = \frac{1}{5}x - \frac{2}{5}$ or $5y = x - 2$

Comments

To solve questions like this,
it is often helpful to draw a
diagram.

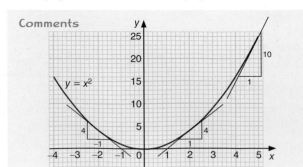

The gradient of the line
$y = {}^-5x + 7$ is ${}^-5$ so the
gradient of the perpendicular
line is the negative reciprocal
of ${}^-5$. The gradient of the
perpendicular line is $\frac{1}{5}$ (since ${}^-5 \times \frac{1}{5} = {}^-1$).

Therefore, this line has the equation $y = \frac{1}{5}x + c$.
Since this line crosses the x-axis at $x = 2$, giving
the point $(2, 0)$, use this to work out the value of c.

$y = \frac{1}{5}x + c$

$0 = \frac{1}{5} \times 2 + c$ Using the fact that when $x = 2$,
           $y = 0$.

$0 = \frac{2}{5} + c$

$c = \frac{-2}{5}$

So the equation of the line is $y = \frac{1}{5}x - \frac{2}{5}$ or
$5y = x - 2$.

Q2 a) gradient $= 4$ **b)** gradient $= {}^-4$
    c) gradient $= 10$ **d)** gradient $= 0$

Comments

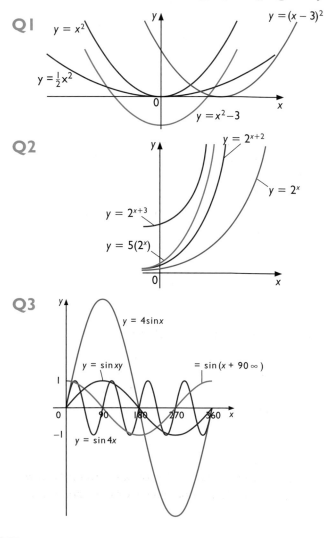

Q3 a) gradient $= 12$ **b)** gradient $= 48$
    c) gradient $= 3$ **d)** gradient $= 27$

Comments

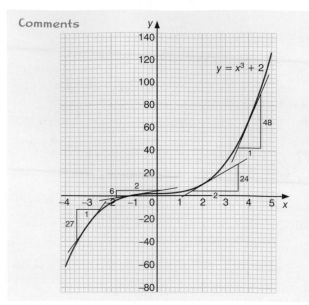

20 Further functions and graphs (page 81)

Q1

Q2

Q3

Unit 3: Shape, Space and Measures

1 Geometric terms (page 87)

Q1 $a = 54°, b = 79°, c = 63°$

> **Comments**
> $a = 54°$ Angles of a triangle add up to 180°.
> $b = 79°$ Angles of a triangle add up to 180°, the base angles of an isosceles triangle are equal.
> $c = 63°$ Angles of a quadrilateral add up to 360°.

Q2 The smallest angle $= 40°$

> **Comments**
> The angles of a triangle add up to 180° so
> $2x + 3x + 4x = 180°$
> $9x = 180°$
> $x = 20°$
> The smallest angle $= 2x = 2 \times 20° = 40°$

Q3 a) Area of triangle $= 27$ cm^2
b) Area of rhombus $= 36.7$ cm^2 (3 s.f.)
c) Area of polygon $= 35$ cm^2

> **Comments**
> **a)** Area of triangle
> $= \frac{1}{2} \times$ base $\times$ perpendicular height
> $= \frac{1}{2} \times 9 \times 6 = 27$ cm^2
> **b)** Area of rhombus = base $\times$ perpendicular height
> $= 7.2 \times 5.1$ Perpendicular height is 5.1 cm
> $\qquad\qquad\qquad\qquad$ (not 7.2 cm).
> $= 36.72 = 36.7$ cm^2 (3 s.f.)
> Rounding to an appropriate degree of accuracy.
> **c)** Area of polygon
> $=$ area of rectangle $+$ area of triangle
> $= 8 \times 3.5 + \frac{1}{2} \times 3.5 \times 4$
> $\qquad$ As length of rectangle $= 12 - 4 = 8$ cm.
> $= 28 + 7 = 35$ cm^2

Q4 a) tetrahedron **b)** hexagonal prism

> **Comments**
> Use paper and scissors to cut out nets and make the given solids, in order to check your answers.

2 Constructions (page 91)

Q1

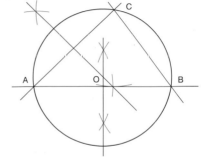

> **Comments**
> Accurate construction will result in the circle **circumscribing** (completely enclosing) triangle ABC and passing through the points A, B and C of the triangle.

Q2

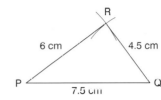

The triangle PQR is right-angled.

> **Comments**
> You should follow the instructions given under the heading 'Constructing a triangle given three sides'.

Q3

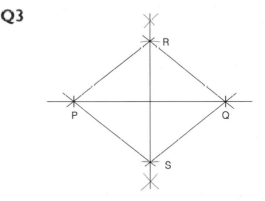

The special name given to this quadrilateral is 'rhombus'.

> **Comments**
> Accurate construction will result in a rhombus PRQS.

3 Maps and scale drawings (page 93)

Q1 a) 270°   **b)** 300°
   c) 165°   **d)** 023°

Comments

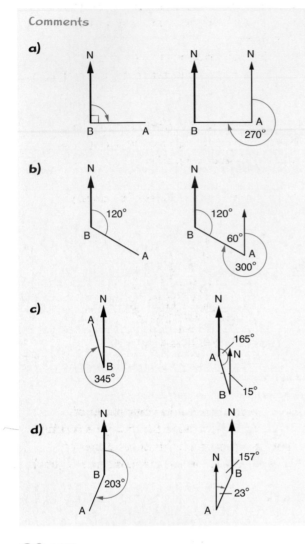

Q2 225°

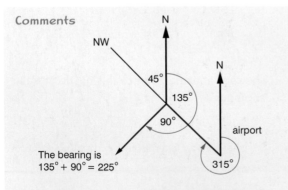

The bearing is
135° + 90° = 225°

A south-west direction is equivalent to a bearing of
225°. As the two north lines are parallel you can
use the fact that interior angles between parallel
lines add up to 180°.

Q3 a) $22\frac{1}{2}$ miles   **b)** 7.2 inches

Comments

a) If 1 inch represents 5 miles then $4\frac{1}{2}$ inches
represents $4\frac{1}{2} \times 5$ miles $= 22\frac{1}{2}$ miles.

b) As 5 miles represents 1 inch then 1 mile
represents $\frac{1}{5}$ inch   Dividing both sides by 5.
36 miles represents $36 \times \frac{1}{5}$ inches $= 7.2$ inches.

Q4 The distance between them is 5.1 miles.

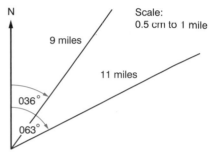

Comments

After two hours the explorers travel 9 miles on a
bearing of 036° and 11 miles on a bearing of 063°.
The distance between them is 2.55 cm which
converts to 5.1 miles, using the given scale.

4 Imperial and metric units (page 95)

Q1 1100 yards

Comments

To answer the question you need to write
kilometres in terms of yards, using
8 kilometres ≈ 5 miles
                 $= 5 \times 1760$ yards (1 mile = 1760 yards)
                 $= 8800$ yards

1 kilometre $= \dfrac{8800}{8}$ Dividing by 8 to find 1 kilometre.
               $= 1100$ yards

There are approximately 1100 yards in one kilometre.

Q2 225 m

Comments

Writing 250 yards $= 750$ feet (3 feet = 1 yard)
                $\approx 750 \times 30$ cm (1 foot ≈ 30 cm)
                $= 22\,500$ cm

                $= \dfrac{22\,500}{100}$ m   Dividing by 100 to
                               convert to metres.

                $= 225$ m

There are approximately 225 metres in 250 yards.

Q3 6.4 kg or 6 kg

Comments
You know that 1 stone = 14 pounds and
2.2 pounds ≈
1 kilogram
So 1 pound ≈ $\frac{1}{2.2}$ kg. Dividing by 2.2 to find 1 pound.
1 stone = 14 pounds
$\approx 14 \times \frac{1}{2.2}$ kg (1 pound ≈ $\frac{1}{2.2}$ kg)
$= 6.363\ 636...$ kg
$= 6.4$ kg or 6 kg
Rounding to an appropriate degree of
accuracy.
There are approximately 6.4 kg or 6 kg in 1 stone.

5 Locus of points (page 97)

Q1

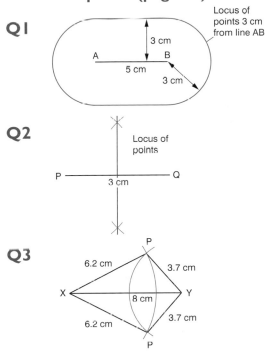

Q2

Q3

P is found at the intersection of the two arcs
(two possible solutions)

Q4

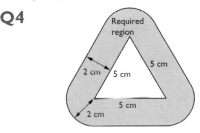

Comments
Sufficient detail is included on the diagrams to
help you complete these accurately.

6 Symmetry (page 99)

Q1
  a) C, D, E, H, I, each have a horizontal line of
  symmetry.
  b) A, H, I each have a vertical line of symmetry.
  c) H, I have both horizontal and vertical lines of
  symmetry.
  d) B, F, G, J, K have no line symmetry.
  e) H, I have rotational symmetry of order 2.

Comments
It is always important to read the question
carefully. The correct types of symmetry must be
identified, for maximum marks.

Q2 A 50p piece has seven lines of symmetry.

Comments

Q3 8
  a) A triangular prism has four planes of
  symmetry (provided the triangle is regular).
  b) A hexagonal prism has seven planes of
  symmetry (provided the hexagon is regular).

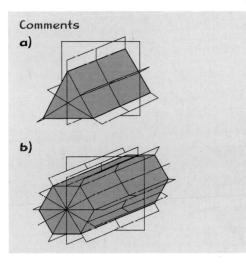

Comments
a)

b)

7 Transformations (page 103)

Q1 The single transformation is a translation of $\begin{pmatrix} ^-5 \\ 3 \end{pmatrix}$.

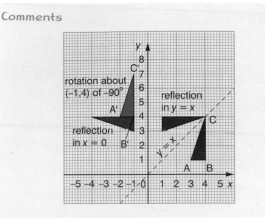

Q2 **a)** $(^-3, 4)$ **b)** $(3, ^-4)$
    c) $(1, 4)$ **d)** $(4, 3)$

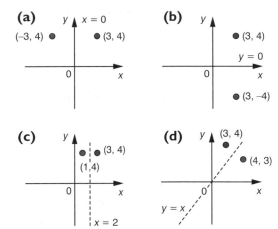

Comments

A reflection is such that the object and image are the same distance away from the given line.

Q3 **a)** (i)   (ii)

    b) (i)   (ii)

    c) (i)   (ii)

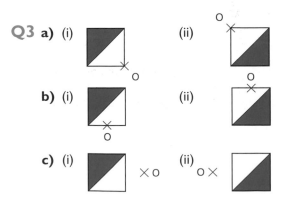

Comments

For the rotation, any two corresponding points on the object and image make the same angle ($^+90°$ in part (i) and $^-90°$ in part (ii)) at the centre of rotation.

Q4 a)   **b)**

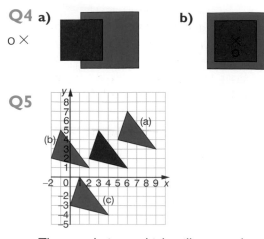

Q5

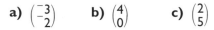

The translations which will return the image to ABC are given by:

a) $\begin{pmatrix} ^-3 \\ ^-2 \end{pmatrix}$ **b)** $\begin{pmatrix} 4 \\ 0 \end{pmatrix}$ **c)** $\begin{pmatrix} 2 \\ 5 \end{pmatrix}$

Comments

After a translation of $\begin{pmatrix} a \\ b \end{pmatrix}$ the translation that will return the image to the object is $\begin{pmatrix} ^-a \\ ^-b \end{pmatrix}$.

Q6

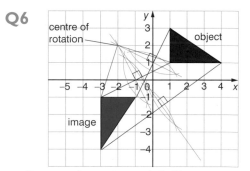

Centre of rotation is $(^-2, 2)$.
Angle of rotation is $^-90°$.

Comments

To find the centre of rotation you should join corresponding points on the object and image with straight lines and draw the perpendicular bisectors of these lines. The centre of rotation lies on the intersection of these perpendicular bisectors. To find the angle of rotation you should join corresponding points on the object and image to the centre of rotation. The angle between these lines is the angle of rotation.

8 Angle properties (page 106)

Q1 $a = 18°$ $b = 117°$
   $c = 63°$ $d = 63°$

Comments

$a = 18°$ Angles on a straight line add up to 180°.
$b = 117°$ Vertically opposite angle.
$c = 63°$ Angles on a straight line add up to 180°.
$d = 63°$ Vertically opposite angle or angles on a straight line add up to 180°.

Q2 Your answer should look something like this.

Let the angles of the triangle ABC be α, β and γ.

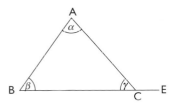

$\alpha + \beta + \gamma = 180°$ Angle sum of a triangle.
$\gamma = 180° - \alpha - \beta$ Rearranging the formula.
$\angle ACE = 180° - \angle ACB$
$\angle ACE = 180° - \gamma$
   $= 180° - (180° - \alpha - \beta)$ From above.
   $= \alpha + \beta$

As $\angle ACE$ is the exterior angle of the triangle, then the exterior angle is equal to the sum of the interior opposite angles of the triangle.

Q3 $a = 117°$ $b = 68°$ $c = 49°$ $d = 87°$
   $e = 87°$ $f = 87°$ $g = 87°$ $h = 93°$

Comments

$a = 117°$ Interior angles between parallel lines add up to 180°. $(180° - 63° = 117°)$
$b = 68°$ Interior angles between parallel lines add up to 180°. $(180° - 112° = 68°)$
$c = 49°$ Angles in a triangle add up to 180°. $(180° - 63° - 68° = 49°)$
$d = 87°$ Corresponding angles between parallel lines.
$e = 87°$ Corresponding angles between parallel lines (the other pair).
$f = 87°$ Vertically opposite angle to angle e.
$g = 87°$ Alternate angle between parallel lines with angle f.
$h = 93°$ Angles on a straight line add up to 180°.

Q4 Each interior angle = 144°

Comments

A decagon is a ten-sided shape.
Use the formula.
Angle sum of an n-sided polygon $= (n - 2) \times 180°$
Angle sum of a ten-sided polygon $= (10 - 2) \times 180°$
   $= 8 \times 180°$
   $= 1440°$
As the polygon is a regular polygon then all the interior angles are equal so $1440° \div 10 = 144°$.

9 Congruence and similarity (page 109)

Q1 $\angle PTQ = \angle RTS$ Vertically opposite angles.
   $TQ = TR$ Given.
   $TP = TS$ Given.
   so $\triangle PQT \equiv \triangle SRT$ As two sides and the included angle of one triangle are equal to two sides and the included angle of the other (SAS).
   $RS = 4.2$ cm

Comments

$\triangle PQT \equiv \triangle SRT$
As $\triangle PQT \equiv \triangle SRT$ then RS = PQ = 4.2 cm.

Q2 AC = 15 cm

Comments

$\triangle AXY$ is similar to $\triangle ABC$.
$\angle XAY = \angle BAC$ They are the same angle.
$\angle AXY = \angle ABC$ Corresponding angles between parallel lines.
so $\triangle AXY$ is similar to $\triangle ABC$ As two angles of one triangle are equal to two angles of the other.
As the two triangles are similar then the ratios of the corresponding sides are equal.

So $\dfrac{AX}{AB} = \dfrac{AY}{AC}$

   $\dfrac{8}{12} = \dfrac{10}{AC}$

   $\dfrac{12}{8} = \dfrac{AC}{10}$ Turning both sides upside-down.

   $AC = \dfrac{12}{8} \times 10$ Multiplying both sides by 10.
   $AC = 15$ cm

Q3 MN = 7.2 cm

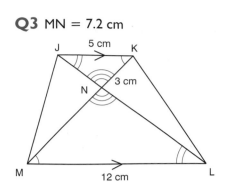

Comments

Drawing a diagram helps to see the situation clearly.

∠JNK = ∠LNM   Vertically opposite angles.

∠KJL = ∠JLM   Alternate angles between parallel lines.

So ΔLMN is similar to ΔJKN as two angles of one triangle are equal to two angles of the other.

As the two triangles are similar then the ratios of the corresponding sides are equal.

So $\dfrac{JK}{ML} = \dfrac{NK}{MN}$

$\dfrac{5}{12} = \dfrac{3}{MN}$

$\dfrac{12}{5} = \dfrac{MN}{3}$ Turning both sides upside-down.

$MN = \dfrac{12}{5} \times 3$ Multiplying both sides by 3.

MN = 7.2 cm

10 Pythagoras' theorem in two dimensions (page 111)

Q1

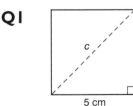

The length of a diagonal = √50 cm

Comments

Drawing a diagram helps to see the situation.
Using Pythagoras' theorem:
$a^2 + b^2 = c^2$
$5^2 + 5^2 = c^2$ Substituting values for a and b.
$c^2 = 50$ Squaring 5 and adding.
$c = \sqrt{50}$ Taking square roots to find c.
The length of a diagonal = √50 cm

Q2 Area of triangle
= 15.6 cm² (3 s.f.)

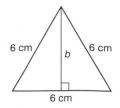

Comments

The triangle can be split into two right-angled triangles to find the height. Drawing a diagram helps to see the situation.
Using Pythagoras' theorem:
$a^2 + b^2 = c^2$
$3^2 + b^2 = 6^2$ Length of the hypotenuse is 6 cm.
$9 + b^2 = 36$ Squaring individual lengths.
$b^2 = 36 - 9$ Making the height the subject.
$b^2 = 27$
$b = 5.196\ 152\ 4$ Taking square roots to find b.
Area of triangle = $\frac{1}{2}$ × base × perpendicular height
         = $\frac{1}{2}$ × 6 × 5.196 152 4
         = 15.588 457
         = 15.6 cm² (3 s.f.)

Q3 Distance = 29.2 km (3 s.f.)

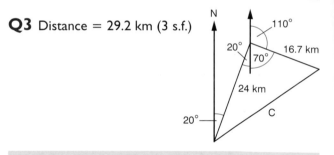

Comments

Drawing a diagram helps to see the situation.
Since the diagram includes a right-angled triangle you can use Pythagoras' theorem.
$a^2 + b^2 = c^2$
$16.7^2 + 24^2 = c^2$ Substituting values for a and b.
$c^2 = 854.89$ Squaring and adding.
$c = 29.238\ 502$ Taking square roots to find c.
Distance = 29.2 km (3 s.f.)

11 Sine, cosine and tangent in right-angled triangles (page 116)

Q1 **a)** a = 6.16 cm (3 s.f.) **b)** b = 9.15 cm (3 s.f.)
    c) c = 3.00 cm (3 s.f.)

Comments

a) $\sin 38° = \dfrac{a}{10}$ a = 6.16 cm (3 s.f.)

b) $\cos 49° = \dfrac{6}{b}$ $b = \dfrac{6}{\cos 49°}$ = 9.15 cm (3 s.f.)

c) $\tan 15° = \dfrac{c}{11.2}$ c = 3.00 cm (3 s.f.)

Q2 a) $a = 40.6°$ (3 s.f.)   **b)** $b = 42.8°$ (3 s.f.)
c) $c = 52.3°$ (3 s.f.)

Comments

a) $\tan a = \dfrac{6}{7}$ $a = \tan^{-1}\dfrac{6}{7}$

$a = \tan^{-1} 0.857\,142\,8... = 40.6°$ (3 s.f.)

b) $\cos b = \dfrac{11}{15}$ $b = \cos^{-1}\dfrac{11}{15}$

$b = \cos^{-1} 0.733\,333\,3... = 42.8°$ (3 s.f.)

c) $\sin c = \dfrac{3.8}{4.8}$ Converting 48 mm to 4.8 cm.

$c = \sin^{-1}\dfrac{3.8}{4.8} = \sin^{-1} 0.791\,666\,6...$

$c = 52.3°$ (3 s.f.)

Q3 $39.8°$ (3 s.f.)

Comments

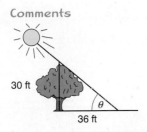

Let the required angle be θ.

$\tan \theta = \dfrac{30}{36}$

$\theta = 39.805\,571° = 39.8°$ (3 s.f.)

12 Lengths, areas and volumes (page 119)

Q1 a)

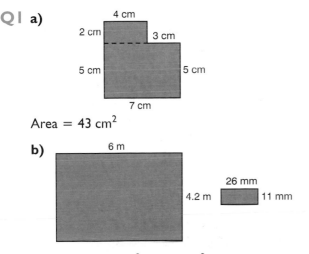

Area = 43 cm^2

b)

Area = $25\,200$ cm^2 or 25.2 m^2 (3 s.f.)

Comments

a) Drawing a diagram helps to see the situation clearly.

Area = $4 \times 2 + 5 \times 7 = 8 + 35 = 43$ cm^2

b) Drawing a diagram helps to see the situation clearly.

Area = $600 \times 420 - 2.6 \times 1.1$

Writing all lengths in the same units.

$= 252\,000 - 2.86$

$= 251\,997.14$

$= 25\,200$ cm^2 (3 s.f.) or 25.2 m^2 (3 s.f.)

if all units are converted into metres.

Q2

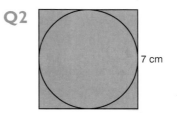

7 cm

Area of waste = 10.5 cm^2
Percentage waste = 21.5% (3 s.f.)

Comments

Drawing a diagram helps to see the situation clearly.

Area of waste = area of square − area of circle

$= 7 \times 7 - \pi \times 3.5^2$

$= 10.515\,49$ cm^2

$= 10.5$ cm^2 (3 s.f.)

Percentage waste = $\dfrac{10.515\,49}{49} \times 100\%$

As original area = 49 cm^2.

$= 21.460\,184\%$

$= 21.5\%$ (3 s.f.)

Q3 Diameter = 22.6 m (3 s.f.)

Comments

Area of pond = πr^2

$\pi r^2 = 400$

$r^2 = \dfrac{400}{\pi}$

$r^2 = 127.323\,95$

$r = \sqrt{127.323\,95}$

$r = 11.283\,792$

$d = 2 \times 11.283\,792$ As diameter = $2 \times$ radius.

$d = 22.567\,583$

$d = 22.6$ m (3 s.f.)

An answer correct to 3 s.f. or 2 s.f. would seem most appropriate in view of the original data.

Q4

6 mm
3 mm

Area of surface of washer = 27π mm^2

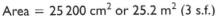

Comments

Area of washer = outside area − inside area

$$= \pi \times 6^2 - \pi \times 3^2$$
$$= 36\pi - 9\pi$$
$$= 27\pi \text{ mm}^2$$

Q5

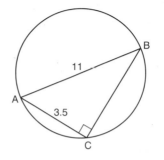

22 mm

3 mm

Volume of the coin $= 363\pi \text{ mm}^3$

Comments

Volume $= \pi r^2 h$ As the shape is a cylinder.

$$= \pi \times 11^2 \times 3 \quad \text{As the radius is 11 mm.}$$
$$= \pi \times 363$$
$$= 363\pi \text{ mm}^3$$

Q6 a) area **b)** length **c)** volume **d)** length

Comments

a) $2\pi r(r + h)$

$r + h$ = length and 2π is a constant, $r(r + h)$ gives units of length × length = area.

b) $\dfrac{\theta}{360} \times 2\pi r$

$\dfrac{\theta}{360} \times 2\pi$ is a constant so r gives units of length.

c) $\dfrac{\pi}{4}d^2 h$

$\dfrac{\pi}{4}$ is a constant and $d^2 h$ gives units of length × length × length = volume.

d) $\sqrt{r^2 + h^2}$

$r^2 + h^2$ gives units of area + area = area. The square root of area = length.

13 Angle properties of circles (page 122)

Q1 $a = 93°$ As opposite angles of a cyclic quadrilateral add up to 180°.

$b = 45°$ As opposite angles of a cyclic quadrilateral add up to 180°.

$c = 90°$ As the angle in a semicircle is 90°.

$d = 90°$ As the angle in a semicircle is 90°.

$e = 96°$ As the angle subtended by an arc at the centre is twice that subtended at the circumference.

$f = 39°$ As the angles subtended by the same arc at the circumference are equal.

$g = 40°$ As vertically opposite angles are equal and the angles of a triangle add up to 180°.

$h = 122°$ As opposite angles of a cyclic quadrilateral add up to 180°.

$i = 90°$ As the angle in a semicircle is 90°.

$j = 32°$ As the angles of a triangle add up to 180°.

$k = 32°$ As j and k are alternate angles between two parallel lines.

Q2 a) BC = 10.4 cm (3 s.f.)

    b) ∠BAC = 71.4° (3 s.f.)

    c) ∠ABC = 18.6° (3 s.f.)

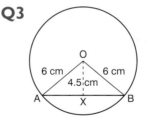

Comments

a) Drawing a diagram helps to see the situation clearly.

∠ACB = 90° As the angle in a semicircle is 90°.

$AB^2 = AC^2 + BC^2$

Applying Pythagoras' theorem to triangle ABC.

$11^2 = 3.5^2 + BC^2$ so BC = 10.428 327 cm

b) Using $\cos\theta = \dfrac{\text{length of adjacent side}}{\text{length of hypotenuse}}$:

$\cos BAC = \dfrac{3.5}{11}$ so ∠BAC = 71.446 995°

c) ∠ABC = 18.553 005°

As the angles of a triangle add up to 180°.

Q3

6 cm   O   6 cm

4.5 cm

A   X   B

AB = 7.94 cm (3 s.f.)

Comments

Drawing a diagram helps to see the situation clearly. Let O be the centre of the circle, OX be the perpendicular bisector of the chord.

For the right-angled triangle OAX:

$6^2 = 4.5^2 + AX^2$ Applying Pythagoras' theorem to the right-angled triangle.

AX = 3.968 627

AB = 2 × AX = 2 × 3.968 627 = 7.937 254

AB = 7.94 cm (3 s.f.)

Q4 a) $\angle POS = 67°$ **b)** $\angle OPR = 23°$
  c) $\angle OPS = 23°$

Comments
a) $\angle POS = 67°$ $\angle POS = \angle POR$ as triangle ORP and
              triangle OSP are congruent (RHS).
b) $\angle OPR = 23°$ $\angle ORP = 90°$ and angles of the
              triangle RPO = 180°.
c) $\angle OPS = 23°$ $\angle OPS = \angle OPR$ as triangle ORP and
              triangle OSP are congruent (RHS).

14 Further areas and volumes (page 125)

Q1 Volume of the sphere is 36π cm^3

Comments
Surface area $= 4\pi r^2 = 36\pi$
$\qquad\quad \pi r^2 = 9\pi$ Dividing both sides by 4.
$\qquad\qquad r^2 = 9$ Dividing both sides by π.
$\qquad\qquad\; r = 3$ Taking the square root
$\qquad\qquad\qquad\quad$ on both sides.
Volume $= \frac{4}{3}\pi r^3$
$\qquad\;\; = \frac{4}{3}\pi \times 3^3$
$\qquad\;\; = 36\pi$ cm^3

Q2 a) Ratio of their surface areas $= 9:25$
  b) Ratio of their volumes $= 27:125$

Comments
If the ratio of corresponding lengths $= 3:5$ then
ratio of corresponding areas $= 3^2:5^2$
ratio of corresponding volumes $= 3^3:5^3$.
a) Ratio of their surface areas $= 3^2:5^2 = 9:25$
b) Ratio of their volumes $= 3^3:5^3 = 27:125$

Q3 Curved surface area $= 10\,485$ mm^2

Comments
If the height of the similar cone is three times that
of the original then the area of the similar cone is
$3^2 (= 9)$ times that of the original.
Curved surface area $= 9 \times 1165$ mm$^2 = 10\,485$ mm^2.

15 Three-dimensional trigonometry (page 127)

Q1 a) $AF = 5\sqrt{5}$ cm

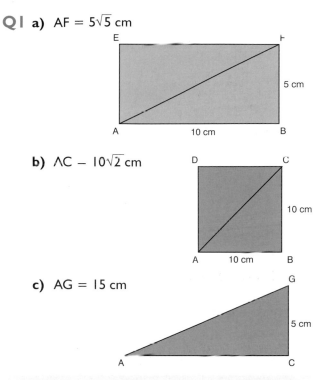

b) $AC = 10\sqrt{2}$ cm

c) $AG = 15$ cm

Comments
a) Drawing a diagram helps to see the situation
    clearly.
    $AF^2 = AB^2 + BF^2$ Using Pythagoras' theorem on
                    the right-angled triangle ABF.
    $AF^2 = 10^2 + 5^2$ As BF = CG.
    $AF^2 = 125$
    $AF = \sqrt{125}$ Taking square roots on both
                    sides.
    $AF = 5\sqrt{5}$ cm In its lowest terms.
b) Drawing a diagram helps to see the situation
    clearly.
    $AC^2 = AB^2 + BC^2$ Using Pythagoras' theorem on
                    the right-angled triangle ABC.
    $AC^2 = 10^2 + 10^2$
    $AC^2 = 200$
    $AC = \sqrt{200}$ Taking square roots on both sides.
    $AC = 10\sqrt{2}$ cm In its lowest terms.
c) Drawing a diagram helps to see the situation
    clearly.
    $AG^2 = AC^2 + CG^2$ Using Pythagoras' theorem on
                    the right-angled triangle ACG.
    $AG^2 = (10\sqrt{2})^2 + 5^2$
    $AG^2 = 200 + 25$
    $AG^2 = 225$
    $AG = \sqrt{225}$ Taking square roots on both sides.
    $AG = 15$ cm

Q2

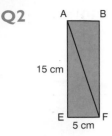

Area of ADGF = $25\sqrt{10}$ cm^2

Comments

Drawing a diagram helps to see the situation clearly.

$AF^2 = AE^2 + EF^2$ — Using Pythagoras' theorem on the right-angled triangle AEF.

$AF^2 = 15^2 + 5^2$

$AF^2 = 250$

$AF = \sqrt{250}$ — Taking square roots on both sides.

$AF = 5\sqrt{10}$ cm

Area of ADGF = $AF \times FG$ — As the plane is a rectangle.

$\quad = 5\sqrt{10} \times 5$

$\quad = 25\sqrt{10}$ cm^2

Q3

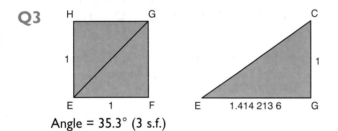

Angle = 35.3° (3 s.f.)

Comments

Drawing a diagram helps to see the situation clearly. Assume the cube to have sides of length one unit (since any other cube will be similar and therefore give the same value for the required angle).

$EG^2 = EF^2 + FG^2$ — Using Pythagoras' theorem on the right-angled triangle EFG.

$EG^2 = 1^2 + 1^2$

$EG^2 = 2$

$EG = 1.414\,2136$ — Taking square roots on both sides.

$\tan CEG = \dfrac{CG}{EG}$

As $\tan\theta = \dfrac{\text{length of opposite side}}{\text{length of adjacent side}}$.

$\tan CEG = \dfrac{1}{1.414\,2136}$

$\tan CEG = 0.707\,1068$

$\angle CEG = 35.264\,390$ — Using $\tan^{-1}$ or arctan to find the angle.

$\angle CEG = 35.3°$ (3 s.f.) — Rounding to an appropriate degree of accuracy.

Q4 WS = 78.0m (3 s.f.)

Comments

To find WS, you need to find WX and XS as follows.

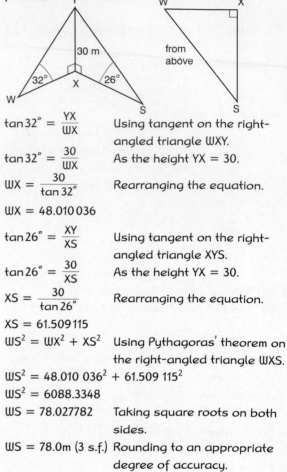

$\tan 32° = \dfrac{YX}{WX}$ — Using tangent on the right-angled triangle WXY.

$\tan 32° = \dfrac{30}{WX}$ — As the height YX = 30.

$WX = \dfrac{30}{\tan 32°}$ — Rearranging the equation.

$WX = 48.010\,036$

$\tan 26° = \dfrac{XY}{XS}$ — Using tangent on the right-angled triangle XYS.

$\tan 26° = \dfrac{30}{XS}$ — As the height YX = 30.

$XS = \dfrac{30}{\tan 26°}$ — Rearranging the equation.

$XS = 61.509\,115$

$WS^2 = WX^2 + XS^2$ — Using Pythagoras' theorem on the right-angled triangle WXS.

$WS^2 = 48.010\,036^2 + 61.509\,115^2$

$WS^2 = 6088.3348$

$WS = 78.027\,782$ — Taking square roots on both sides.

$WS = 78.0m$ (3 s.f.) — Rounding to an appropriate degree of accuracy.

16 Sine, cosine and tangent for angles of any size (page 129)

Q1 **a)** $x = 36.9°$ and $143.1°$ (1 d.p.)

b) $x = 216.9°$ and $323.1°$ (1 d.p.)

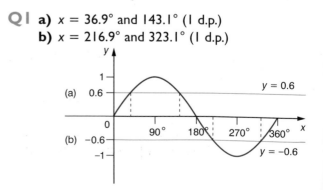

c) $x = 78.5°$ and $281.5°$ (1 d.p.)

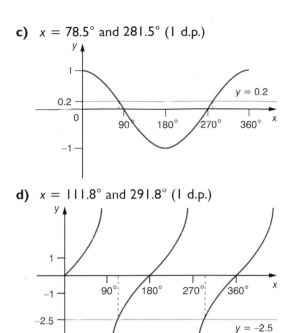

d) $x = 111.8°$ and $291.8°$ (1 d.p.)

Comments

a) Drawing a diagram helps to see the situation clearly.

$x = \sin^{-1} 0.6$, $x = 36.869898$ (from the calculator)

From the graph the solutions in the given range are $x = 36.9°$ and $143.1°$ (1 d.p.).

b) $x = \sin^{-1} {}^-0.6$

$x = {}^-36.869898°$ (from the calculator)

From the graph the solutions in the given range are $216.9°$ and $323.1°$ (1 d.p.).

c) Drawing a diagram helps to see the situation clearly.

$x = \cos^{-1} 0.2$

$x = 78.463041°$ (from the calculator)

From the graph, the solutions in the given range are $78.5°$ and $281.5°$ (1 d.p.).

d) Drawing a diagram helps to see the situation clearly.

$x = \tan^{-1} {}^-2.5$

$x = {}^-68.198591°$ (from the calculator)

From the graph, the solutions in the given range are $111.8°$ and $291.8°$ (1 d.p.).

Q2 a $x = {}^-296.6°, {}^-116.6°, 63.4°$ and $243.4°$ (1 d.p.)

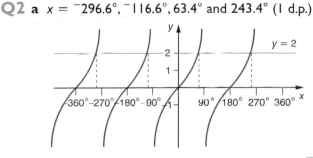

b) $x = {}^-328.3°, {}^-211.7°, 31.7°$ and $148.3°$ (1 d.p.)

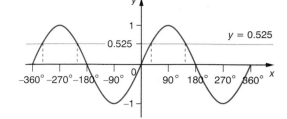

Comments

a) Drawing a diagram helps to see the situation clearly.

Using the calculator:

$\tan x = 2$

$x = \tan^{-1} 2$

$x = 63.434949°$

From the graph, the solutions in the given range are ${}^-296.6°, {}^-116.6°, 63.4°$ and $243.4°$ (1 d.p.).

b) Drawing a diagram helps to see the situation clearly.

Using the calculator:

$\sin x = 0.525$

$x = \sin^{-1} 0.525$

$x = 31.668243°$

From the graph, the solutions in the given range are ${}^-328.3°, {}^-211.7°, 31.7°$ and $148.3°$ (1 d.p.).

Q3 a) $30° \leqslant x \leqslant 150°$

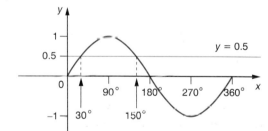

b) $60° < x < 300°$

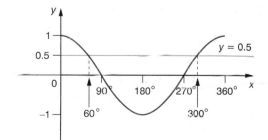

a) Drawing a diagram helps to see the situation
clearly.

$\sin x \geqslant 0.5$

From the graph, x lies in the range $30° \leqslant x \leqslant 150°$.

b) Drawing a diagram helps to see the situation
clearly.

$\cos x < 0.5$

From the graph, x lies in the range $60° < x < 300°$.
Remember that $\cos x < 0.5$ so $x = 60°$ and
$x = 300°$ are not included.

17 Sine and cosine rules (page 132)

Q1 a) $a = 57.6$ mm (3 s.f.)
   b) $B = 70.9°$ (1 d.p.)
   c) $C = 45.8°$ (1 d.p.)
   d) $D = 75.6°$ or $104.4°$ (1 d.p.)

a) Using the cosine rule:

$a^2 = 47^2 + 35^2 - (2 \times 47 \times 35 \times \cos 88°)$

Substituting the given lengths and taking
care with the $-2bc\cos A$ term.

$a = 57.612331$

b) Using the sine rule and both sides.

$\dfrac{a}{\sin A} = \dfrac{b}{\sin B}$ or $\dfrac{\sin A}{a} = \dfrac{\sin B}{b}$

$\dfrac{\sin B}{3.9} = \dfrac{\sin 58°}{3.5}$ Substituting the given values.

$\sin B = 3.9 \times \dfrac{\sin 58°}{3.5}$ Multiplying both sides
by 3.9.

$B = 70.903321°$

c) Using the cosine rule:

$\cos C = \dfrac{4.6^2 + 5.1^2 - 3.8^2}{2 \times 4.6 \times 5.1}$ Substituting the
given lengths.

$C = 45.767605°$

d) Using the sine rule and reciprocating both sides:

$\dfrac{\sin D}{110} = \dfrac{\sin 25°}{48}$ Substituting the given values.

$\sin D = 110 \times \dfrac{\sin 25°}{48}$ Multiplying both sides
by 110.

$\sin D = 0.9685001$

$D = 75.580895°$ or $104.4191°$

As $\sin D$ has two possible solutions here.

Q2 a) Area $= 4.36$ m^2 (3 s.f.)
   b) Area $= 111$ cm^2 (3 s.f.)

a) Area of a triangle $= \frac{1}{2}ab\sin C$
$= \frac{1}{2} \times 3.6 \times 4.7 \times \sin 31° = 4.3572221$

b) Area of a triangle $= \frac{1}{2}ab\sin C$
$= \frac{1}{2} \times 16 \times 16 \times \sin 60°$

As the angles of an equilateral triangle are
all $60°$.

$= 110.85125$

Q3

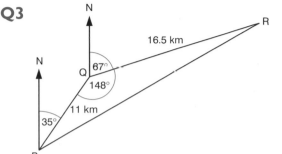

Distance $= 26.5$ km (3 s.f.)
Bearing $= 054°$ (to the nearest whole degree)

By drawing a diagram of the situation you can see
that PR represents the required distance and that
the bearing can be found from the angle QPR.
The bearing of P from Q is $215°$ and the angle
$PQR = 148°$, as $215° - 67° = 148°$.

Using the cosine rule on the triangle PQR:

$a^2 = b^2 + c^2 - 2bc\cos A$

$PR^2 = 11^2 + 16.5^2 - (2 \times 11 \times 16.5 \times \cos 148°)$

Substituting the given lengths and taking care
with the $-2bc\cos A$ term.

$PR = 26.478132$

So the distance $= 26.5$ km (3 s.f.).

To find angle QPR, using the sine rule in the form

$\dfrac{\sin A}{a} = \dfrac{\sin B}{b}$:

$\dfrac{\sin QPR}{QR} = \dfrac{\sin 148°}{PR}$

$\dfrac{\sin QPR}{16.5} = \dfrac{\sin 148°}{26.478132}$ Substituting the given
lengths and $\angle PQR$.

$\sin QPR = 16.5 \times \dfrac{\sin 148°}{26.478132}$ Multiplying both
sides by 16.5.

$\sin QPR = 0.33022223$

$\angle QPR = 19.282265°$

Required bearing $= 35° + 19.282265°$

Bearing $= 35° +$ angle QPR

$= 54.282265°$

$= 054°$ (to the nearest whole
degree)

Remember to write it in the bearing form and
round to an appropriate degree of accuracy.

18 Arc, sector and segment (page 135)

Q1 Arc length $= 6\pi$ cm

Sector area $= 27\pi$ cm^2

> **Comments**
>
> Arc length $= \dfrac{120}{360} \times 2 \times \pi \times 9$
>
> $\qquad = 6\pi$ cm
>
> Sector area $= \dfrac{120}{360} \times \pi \times 9^2$
>
> $\qquad = 27\pi$ cm^2

Q2 Area $= 74.3$ cm^2

> **Comments**
>
> Area of segment = area of sector − area of triangle
>
> Area of sector $= \dfrac{120}{360} \times \pi r^2 = \dfrac{120}{360} \times \pi \times 11^2$
>
> $\qquad = 126.7109$
>
> Area of triangle $= \dfrac{1}{2} ab \sin\theta$
>
> $\qquad = \dfrac{1}{2} \times 11 \times 11 \times \sin 120°$
>
> Where a and b are equal to the radius of the circle.
>
> $\qquad = 52.394537$
>
> Area of segment = area of sector AOB − area of $\triangle$AOB
>
> $\qquad = 126.7109 - 52.394537$
>
> $\qquad = 74.316363$

Q3

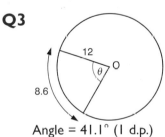

Angle $= 41.1°$ (1 d.p.)

> **Comments**
>
> Drawing a diagram helps to see the situation clearly.
>
> Arc length $= \dfrac{\theta}{360} \times 2\pi r$
>
> $8.6 = \dfrac{\theta}{360} \times 2 \times \pi \times 12$
>
> $\theta = 41.061975°$
>
> Angle subtended by arc $= 41.1°$ (3 s.f.)

Q4

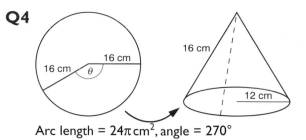

Arc length $= 24\pi$ cm^2, angle $= 270°$

> **Comments**
>
> Drawing a diagram helps to see the situation clearly.
>
> From the diagram you can see that:
>
> **i)** the radius of the circle is the same as the slant height of the cone (given as 16 cm)
>
> **ii)** the circumference of the base of the cone is the same as the arc cut out from the circle.
>
> Circumference of the base of the cone
>
> $= 2\pi r$
>
> $= 2 \times \pi \times 12$
>
> $= 24\pi$
>
> Arc length $= \dfrac{\theta}{360} \times 2\pi r$
>
> $24\pi = \dfrac{\theta}{360} \times 2 \times \pi \times 16$ As arc length = circumference of base of cone.
>
> $\theta = 270°$ Cancelling π on both sides and working out.

19 Further angle properties of circles (page 137)

Q1

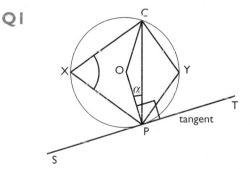

Let $\angle OPC = \alpha$

Then $\angle OCP = \alpha$ Base angles of isosceles triangle.

and $\angle COP = 180° - 2\alpha$ Angle sum of a triangle.

$\angle CXP = \dfrac{180° - 2\alpha}{2}$ The angle subtended by the chord PC at the centre is twice the angle subtended at the circumference.

$\angle CXP = 90° - \alpha$ Simplify the right-hand side.

∠OPT = 90°

A tangent to the circle is perpendicular to the radius at the point of contact.

∠CPT = 90° − α

As ∠OPT = 90° and ∠OPC = α.

Since ∠CPT = 90° − α and ∠CXP = 90° − α then ∠CPT = ∠CXP.

Comments

Drawing a diagram helps to see the situation clearly. Add the radii OC and OP.

Q2 a) ∠PXC = 61°
   b) ∠PYC = 61°

Comments

a) ∠PXC = 61°   ∠PXC = ∠CPT by the alternate segment theorem.

b) ∠PYC = 61°   ∠PYC = ∠PXC as these are angles subtended by the same arc PC.

Q3 a) ∠COP = 76°
   b) ∠CPT = 38°
   c) ∠OCP = 52°

Comments

a) ∠COP = 76°   The angle subtended by the arc PC at the centre is twice that subtended at the circumference.

b) ∠CPT = 38°   ∠CPT = ∠CDP by the alternate segment theorem.

c) ∠OCP = 52°   ∠OCP = ∠OPC as they are base angles of the isosceles triangle COP where ∠COP = 76°.

20 Enlargement with negative scale factors (page 139)

Q1

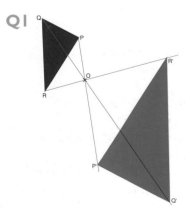

Comments

An enlargement of scale factor ⁻2 will double the length of all of the sides but the enlargement will be on the opposite side of the centre of enlargement.

Q2 The single transformation is a rotation of 180° about the origin.

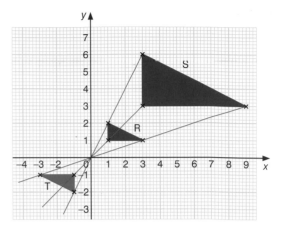

Comments

The enlargement of scale factor $-\frac{1}{3}$ has the effect of producing an enlargement where the new lengths are $\frac{1}{3}$ of the original lengths. The negative enlargement means that the enlargement occurs on the opposite side of the centre of enlargement.

Q3 Centre (0, 1), enlargement $-\frac{1}{2}$

Comments

The centre of enlargement is (0, 1).

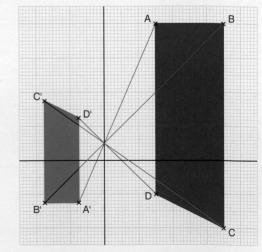

Corresponding lengths on A'B'C'D' are half of those on ABCD and the position of the enlargement on the opposite side of the centre suggest an enlargement scale factor $-\frac{1}{2}$.

21 Vectors and vector properties (page 142)

Q1 a) $\vec{AC} = 2\mathbf{a}$ **b)** $\vec{AM} = 3\mathbf{b}$

   c) $\vec{AF} = \mathbf{a} + \mathbf{b}$ **d)** $\vec{AK} = 2\mathbf{a} + 2\mathbf{b}$

   e) $\vec{GA} = {}^-2\mathbf{a} - \mathbf{b}$ **f)** $\vec{PE} = {}^-3\mathbf{a} - 2\mathbf{b}$

Comments

Remember that $\mathbf{a} + \mathbf{b} = \mathbf{b} + \mathbf{a}$
and if $\vec{AB} = \mathbf{a}$ then $\vec{BA} = {}^-\mathbf{a}$.

Q2 a) $|\vec{AB}| = 5$ units   **b)** $|\vec{BC}| = \sqrt{26}$ units

   c) $|\vec{AC}| = \sqrt{73}$ units

Comments

Using the fact that $|\vec{AB}| = \sqrt{x^2 + y^2}$

a) $|\vec{AB}| = \left|\begin{pmatrix} 3 \\ 4 \end{pmatrix}\right| = \sqrt{3^2 + 4^2} = \sqrt{25} = 5$

b) $|\vec{BC}| = \left|\begin{pmatrix} 5 \\ -1 \end{pmatrix}\right| = \sqrt{5^2 + (^-1)^2} = \sqrt{26}$

c) $|\vec{AC}| = \left|\begin{pmatrix} 8 \\ 3 \end{pmatrix}\right| = \sqrt{8^2 + 3^2} = \sqrt{73}$

Q3 a) $\vec{AC} = \mathbf{a} + \mathbf{b}$ **b)** $\vec{AO} = \mathbf{b}$

   c) $\vec{OB} = \mathbf{a} - \mathbf{b}$ **d)** $\vec{AD} = 2\mathbf{b}$

ACDF is a parallelogram.

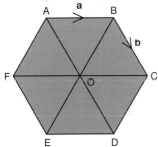

Comments

a) $\vec{AC} = \vec{AB} + \vec{BC} = \mathbf{a} + \mathbf{b}$

b) $\vec{AO} = \mathbf{b}$

c) $\vec{OB} = \vec{OA} + \vec{AB} = {}^-\mathbf{b} + \mathbf{a} = \mathbf{a} - \mathbf{b}$

d) $\vec{AD} = 2\mathbf{b}$

   $\vec{AC} = \mathbf{a} + \mathbf{b}$ and $\vec{FD} = \mathbf{a} + \mathbf{b}$

   $\vec{AF} = \mathbf{b} - \mathbf{a}$ and $\vec{CD} = \mathbf{b} - \mathbf{a}$

   so AC is parallel and equal to FD

   AF is parallel and equal to CD.

   Therefore ACDF is a parallelogram.

UNIT 4: HANDLING DATA
1 Collecting and representing data (page 148)

Q1 a)

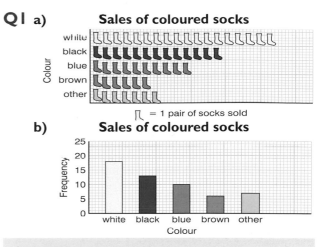

Comments

Remember to give your representations a title and keep the symbols simple.

Q2 Heights of plants (cm)

3	2	3	7	8
4	0	2	6	9	9
5	0	1	1		

Key: 3|3 means 3.2

Comments

Remember the key and the title.

Q3

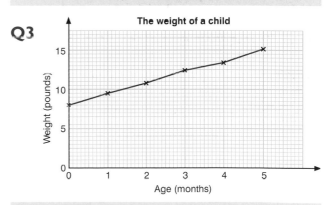

Comments

a) Underground (from the pie chart)

b) Walking (from the pie chart)

c) $10\% \times 360° = 36°$

   The number travelling by bus or by
   car $= 45\% \times 100 = 45$ people

   Number who travelled by bus is twice the
   number who travelled by car.

d) Travelled by bus: $\frac{2}{3} \times 45 = 30$

e) Travelled by car: $\frac{1}{3} \times 45 = 15$

Q4

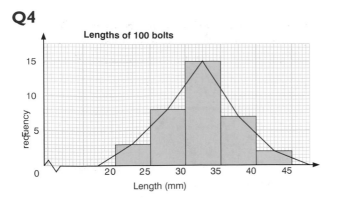

Lengths of 100 bolts

Comments

Remember that the lines at each end should be extended to the horizontal axis as shown.

Q5

Courses followed by 250 students

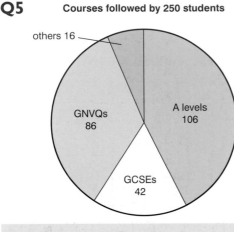

others 16

GNVQs 86

A levels 106

GCSEs 42

Comments

The pie chart needs to be drawn to represent 250 people. There are 360° in a full circle so each person is represented by 360° ÷ 250 = 1.44° of the pie chart.

Course	Number	Angle
A levels	106	106 × 1.44° = 153°
GCSEs	42	42 × 1.44° = 60°
GNVQs	86	86 × 1.44° = 124°
Others	16	16 × 1.44° = 23°
		360°

All angles are rounded to the nearest degree. The angles do not work out exactly, so they have to be rounded to the nearest whole number.

Q6 a) underground **b)** walking **c)** 36°
d) 30 **e)** 15

2 Measures of central tendency (page 153)

Q1 a) Science fiction books
b) Winchester House
c) Apples

Comments

a) Science Fiction books – greatest number of pictures
b) Winchester House – longest bar
c) Apples – largest angle/sector area

Q2 Median = 2

Comments

To find the median of a frequency distribution you need to find the cumulative frequency.

Number of goals	Frequency	Cumulative frequency
0	4	4
1	8	12
2	11	23
3	7	30
4	3	33
5	0	33
6	1	34

Median value = $\frac{1}{2}(34 + 1) = 17\frac{1}{2}$th value which is 2.

Q3 17.6 years (3 s.f.)

Comments

Age (years) x	Frequency f	Frequency × age fx
17	23	391
18	13	234
19	4	76
20	0	0
21	1	21
	$\Sigma f = 41$	$\Sigma fx = 722$

Mean = $\frac{\Sigma fx}{\Sigma f} = \frac{722}{41} = 17.609756$

= 17.6 years (3 s.f.)

Q4 23 kg to an appropriate degree of accuracy.

Comments

Weight (kg)	Mid-interval value x	Frequency f	Frequency × mid-interval value fx
0–10	5	11	55
10–20	15	18	270
20–30	25	16	400
30–40	35	11	385
40–50	45	5	225
50–60	55	2	110
		$\Sigma f = 63$	$\Sigma fx = 1445$

For the group frequency distribution:

$$\text{mean} = \frac{\Sigma fx}{\Sigma f} = \frac{1445}{63} = 22.936\,508$$

$$= 23 \text{ kg to an appropriate degree of accuracy.}$$

Q5 £9.61 to an appropriate degree of accuracy.

Comments

Amount (£)	Mid-interval value x	Frequency f	Frequency × mid-interval value fx
0 and less than 5	2.5	12	30
5 and less than 10	7.5	15	112.5
10 and less than 15	12.5	8	100
15 and less than 20	17.5	7	122.5
20 and less than 25	22.5	3	67.5
		$\Sigma f = 45$	$\Sigma fx = 432.5$

For the group frequency distribution:

$$\text{mean} = \frac{\Sigma fx}{\Sigma f} = \frac{432.5}{45} = 9.611\,111\,1$$

$$= £9.61 \text{ to an appropriate degree of accuracy.}$$

3 Measures of spread and cumulative frequency diagrams (page 158)

Q1 Range = 33 cm

Interquartile range = 23 cm

Comments

Arranging the information in order:

6 cm, 12 cm, 13 cm, 13 cm, 15 cm, 18 cm, 18 cm, 18 cm, 20 cm, 21 cm, 21 cm, 36 cm, 37 cm, 37 cm, 39 cm

The range of the heights is 39 − 6 = 33 cm.

To find the interquartile range you need to find the lower quartile and the upper quartile. The lower quartile is the $\frac{1}{4}(15 + 1)$ = 4th value and the upper quartile is the $\frac{3}{4}(15 + 1)$ = 12th value.

From the data:

6 cm, 12 cm, 13 cm, 13 cm, 15 cm, 18 cm, 18 cm, 18 cm, 20 cm, 21 cm, 21 cm, 36 cm, 37 cm, 37 cm, 39 cm
         ↑   ↑   ↑
         LQ   Median   UQ

Interquartile range = upper quartile − lower quartile

$$= 36 − 13 = 23 \text{ cm}$$

Q2 a) The range = £34   **b)** IQR – £17

Comments

a) The range = 76 − 42 (from the box plot) = £34

b) IQR = 68 − 51 (from the box plot) = £17

Q3

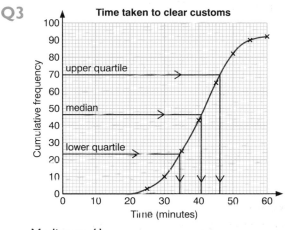

Time taken to clear customs

Median = 41

Interquartile range = $11\frac{1}{2}$

Comments

Time (t minutes)	Frequency	Cumulative frequency
$20 \leqslant t < 25$	3	3
$25 \leqslant t < 30$	7	10
$30 \leqslant t < 35$	15	25
$35 \leqslant t < 40$	18	43
$40 \leqslant t < 45$	22	65
$45 \leqslant t < 50$	17	82
$50 \leqslant t < 55$	8	90
$55 \leqslant t < 60$	2	92

The cumulative frequencies must be plotted at the upper class boundaries (i.e. 25, 30, 35, 40, etc.)

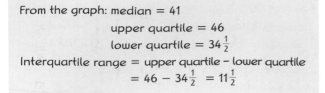

From the graph: median = 41

upper quartile = 46

lower quartile = $34\frac{1}{2}$

Interquartile range = upper quartile − lower quartile

$= 46 - 34\frac{1}{2} = 11\frac{1}{2}$

Q4

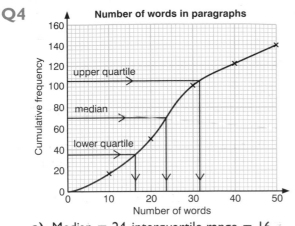

a) Median = 24, interquartile range = 16

b) Percentage of paragraphs over 35 words = 19%

Comments

Number of words per paragraph	Number of paragraphs	Cumulative frequency
1–10	17	17
11–20	33	50
21–30	51	101
31–40	21	122
41–50	18	140

The cumulative frequencies must be plotted at the upper class boundaries (i.e. 10, 20, 30, etc.) as the number of words is discrete in this instance.

a) From the graph: median = 24

upper quartile = 32

lower quartile = 16

Interquartile range

= upper quartile − lower quartile

= 32 − 16 = 16

b) Number of paragraphs under 35 words in length = 114

Number of paragraphs over 35 words in length

= 140 − 114 = 26

Percentage of paragraphs over 35 words in length

$= \frac{26}{140} \times 100 = 19\%$

4 Scatter diagrams and lines of best fit (page 161)

Q1 a) Strong positive correlation

b) Moderate negative correlation

c) No correlation

Q2 a)

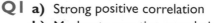

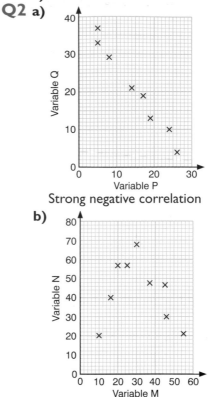

Strong negative correlation

b)

There is no definable correlation between the two variables although there is some correlation here as demonstrated by the U-shaped curve.

Q3

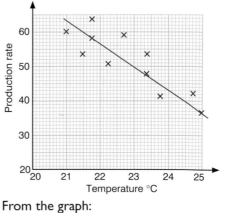

From the graph:

a) production rate = 51 units

b) temperature = 22.9°C

Comments

From the graph:

a) production rate = 51 units, by reading off the data when the temperature equals 22.8°C

b) temperature = 22.9°C, by reading off the data when the production rate equals 50 units.

Q4

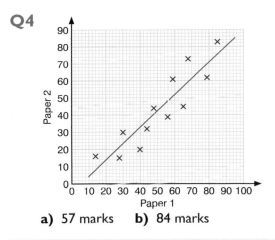

Paper 2 (y-axis) vs **Paper 1** (x-axis)

a) 57 marks   b) 84 marks

Comments

From the graph:

a) 57 marks, by reading off the data when paper 1 equals 65

b) 84 marks, by reading off the data when paper 2 equals 75.

5 Probability (page 164)

Q1 a) Number of blue balls = 16

b) Number of red balls = 23

c) Number of green balls = 11

Comments

a) Number of blue balls = 32% × 50 = 16

b) Number of red balls = 0.46 × 50 = 23

c) Number of green balls = 50 − (16 + 23)

$\qquad$ = 11 As the remaining balls are green.

Q2 p(late) = 0.35

Comments

p(late) = 1 − p(not late)

$\qquad$ = 1 − (0.2 + 0.45) Total probability = 1.

$\qquad$ = 0.35

Q3

		Die					
		1	2	3	4	5	6
Coin	H	H1	H2	H3	H4	H5	H6
	T	T1	T2	T3	T4	T5	T6

a) $\frac{1}{12}$

b) $\frac{1}{4}$

Comments

From the possibility space diagram, there are 12 possible outcomes.

a) p(a head and a one) = $\frac{1}{12}$ As only one outcome gives this result.

b) p(a tail and an odd number)

$= \frac{3}{12}$ As three outcomes give this result.

$= \frac{1}{4}$ Cancelling down to its lowest terms.

Q4

		Second die			
		1	2	3	4
	1	2	3	4	5
	2	3	4	5	6
First die	3	4	5	6	7
	4	5	6	7	8

a) $\frac{1}{16}$ b) $\frac{3}{16}$ c) 0

The most likely outcome is a 5.

Comments

From the possibility space diagram, there are 16 possible outcomes.

a) p(total of 2) = $\frac{1}{16}$ As only one outcome gives this result.

b) p(total of 6) = $\frac{3}{16}$ As three outcomes give this result.

c) p(total of 9) = 0 As there is no possibility of getting a total of 9.

The most likely outcome is a 5.

Q5 40

Comments

p(number greater than 4) = $\frac{2}{6} = \frac{1}{3}$

Expected frequency = $120 \times \frac{1}{3} = 40$

Q6 1 car (rounding to an appropriate degree of accuracy)

Comments

Expected number of cars

$$= 1037 \times 0.062\% = 1037 \times \frac{0.062}{100}$$

$$= 0.642\ 94$$

$$= 1\ car\ (rounding\ to\ an\ appropriate\ degree\ of\ accuracy)$$

6 The addition and multiplication rules (page 167)

Q1 **a)** $\frac{1}{3}$ **b)** $\frac{1}{2}$ **c)** $\frac{1}{2}$

Comments

a) $p(5\ or\ 6) = p(5) + p(6)$

As the events are mutually exclusive.

$$= \frac{1}{6} + \frac{1}{6} = \frac{2}{6} = \frac{1}{3}$$

b) $p(even\ number)$

$$= p(2\ or\ 4\ or\ 6)$$

As the events are mutually exclusive.

$$= p(2) + p(4) + p(6)$$

$$= \frac{1}{6} + \frac{1}{6} + \frac{1}{6} = \frac{3}{6} = \frac{1}{2}$$

c) $p(factor\ of\ 8) = p(1\ or\ 2\ or\ 4)$

As the factors of 8 are 1, 2 4 and 8.

$$= p(1) + p(2) + p(4)$$

As the events are mutually exclusive.

$$= \frac{1}{6} + \frac{1}{6} + \frac{1}{6} = \frac{3}{6} = \frac{1}{2}$$

Q2 **a)** $\frac{1}{11}$ **b)** $\frac{2}{11}$ **c)** $\frac{4}{11}$

Comments

a) $p(letter\ P) = \frac{1}{11}$

b) $p(letter\ B) = \frac{2}{11}$

c) $p(letter\ B\ or\ letter\ I) = p(letter\ B) + p(letter\ I)$

$$= \frac{2}{11} + \frac{2}{11} = \frac{4}{11}$$

As the events are mutually exclusive.

Q3

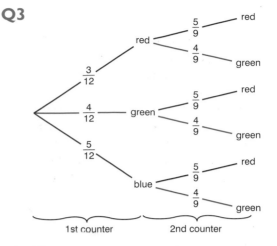

Comments

See the tree diagram.

Q4 **a)** 0.176   **b)** 0.144

Comments

a) $p(lights\ and\ brakes)$

$$= p(lights) \times p(brakes)$$ As the events are independent.

$$= 0.32 \times 0.55 = 0.176$$

b) The probability that the car fails because of its lights only is equal to the probability that the car fails because of its lights and does not fail because of the brakes.

The probability that a car will fail its MOT because of the brakes is 0.55.

The probability that a car will not fail its MOT because of the brakes is 0.45.   (1 – 0.55)

$p(car\ fails\ because\ of\ its\ lights\ only)$

$$= 0.32 \times 0.45 = 0.144$$

Q5 **a)** 0.000 225 or 2.25×10^{-4}

b) 0.029 55

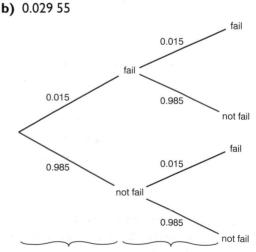

Comments

a) p(both components will fail) = p(fail and fail)
= p (fail) × p(fail)   As the events
are independent.
= 0.015 × 0.015
= 0.000 225 or 2.25×10^{-4}

b) The probability that exactly one component
will fail is equivalent to the probability that the
first component fails and the second doesn't
or the first component doesn't and the second
fails.
= p(fails and doesn't fail or doesn't fail and
fails)
= p(fails and doesn't fail) + p(doesn't fail and
fails) As the events are mutually exclusive.
= p(fails) × p(doesn't fail)
  + p(doesn't fail) × p(fails)
As the events are independent.
= 0.015 × 0.985 + 0.985 × 0.015
As p(doesn't fail) = 1 − p(fail)
= 0.029 55

7 Sampling methods (page 169)

Q1 Choose three from:
(i) Only people with access to a telephone
will be contacted.
or (ii) The survey is limited to only one town.
or (iii) Some of the respondents may not have
used the bus 'last week'.
or (iv) Some of the respondents may be at work
during the evening.

Comments
Responses need to take account of the people
surveyed and the question asked. Here, the words
'travel on local buses', 'telephoned 100 people', 'in
a certain town' and 'one evening' are all relevant.

Q2 a) The villages have different sized populations and
stratified sampling would be most appropriate.
b) Sample size:

Atford	48
Beeham	79
Calbridge	113

Comments
Population of all villages = 12 500
For stratified sampling, the following sample sizes
are appropriate.

Atford = $\frac{2500}{12\,500} \times 240 = 48$

Beeham = $\frac{4100}{12\,500} \times 240 = 78.72 = 79$

Calbridge = $\frac{5900}{12\,500} \times 240 = 113.28 = 113$

Answers for Beeham and Calbridge are rounded
to the nearest integer to give sensible numbers.

Q3 Allows you to identify problems and to improve
the design of the actual survey.

Q4 a) The question suffers from bias in that it
suggests the magazine is 'improved'.
b) The question is unclear, as there is no
mention of a time scale.
c) The question is confusing to answer.

Comments
A good questionnaire should be short, simple, clear
and precise. The questions should be unambiguous,
written in appropriate language, free from bias, and
with no personal or offensive questions.

8 Histograms – unequal class intervals (page 173)
Q1

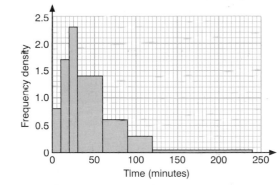

Comments
Completing the table:

Time (t minutes)	Frequency	Class width	Frequency density
0 ≤ t < 10	8	10	0.8
10 ≤ t < 20	17	10	1.7
20 ≤ t < 30	23	10	2.3
30 ≤ t < 60	42	30	1.4
60 ≤ t < 90	18	30	0.6
90 ≤ t < 120	9	30	0.3
120 ≤ t < 240	3	120	0.025

The histogram is shown in the answer.

Q2

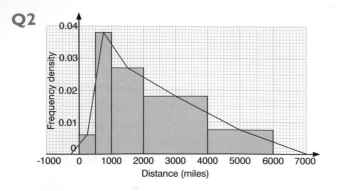

Frequency density / Distance (miles)

Comments

Completing the table:

Distance (miles)	Frequency	Class width	Frequency density
0–500	3	500.5	0.005 994
501–1000	19	500	0.038
1001–2000	27	1000	0.027
2001–4000	36	2000	0.018
4001–6000	15	2000	0.0075

The histogram and frequency polygon are shown in the answer.

Note: The boundaries on the first interval are 0 and 500.5 so the class width = 500.5.

Q3

Distance (miles)	0–	5–	10–	15–	20–	30–50
Number of lecturers	1	8	13	10	12	6

You should check that the total number of lectures is 50.

Comments

Reading from the histogram and working backwards:

Distance (miles)	0–	5–	10–	15–	20–	30–50
Class width	5	5	5	5	10	20
Frequency density	0.2	1.6	2.6	2.0	1.2	0.3
Number of lecturers	1	8	13	10	12	6

where number of lectures = class width × frequency density.

---

9 Multiplication rule for dependent events (page 175)

Q1

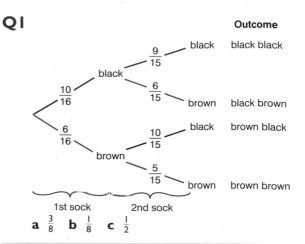

a $\frac{3}{8}$ **b** $\frac{1}{8}$ **c** $\frac{1}{2}$

Comments

A diagram helps to see the situation clearly.

a) p(black and black) $= \frac{10}{16} \times \frac{9}{15} = \frac{3}{8}$

b) p(brown and brown) $= \frac{6}{16} \times \frac{5}{15} = \frac{1}{8}$

c) Reading from the tree diagram p(socks are different colours)

$=$ p(black and brown or brown and black)

$= \frac{10}{16} \times \frac{6}{15} + \frac{6}{16} \times \frac{10}{15} = \frac{1}{2}$

Q2 Probability $= \frac{1}{64}$

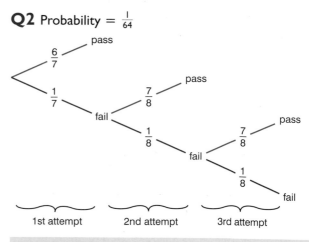

Comments

p(passing on third attempt)

$=$ p(fail and fail and pass)

$= \frac{1}{7} \times \frac{1}{8} \times \frac{7}{8} = \frac{1}{64}$

ANSWERS TO EXAM PRACTICE QUESTIONS

NON-CALCULATOR PAPER

1 $3\frac{8}{15}$

> **Comments**
> $5\frac{1}{5} - 1\frac{2}{3} = \frac{26}{5} - \frac{5}{3}$ Converting to improper fractions.
> $= \frac{78}{15} - \frac{25}{15}$ Writing as equivalent fractions with a common denominator.
> $= \frac{53}{15} = 3\frac{8}{15}$ Rewriting as a mixed number.

2 $8 \times 10^1 \, \text{km}^2$ (square kilometres)

> **Comments**
> Area per head of population $= \frac{4 \times 10^{11}}{5 \times 10^9} = 0.8 \times 10^2$
> $= 8 \times 10^1 \, \text{km}^2$, leaving the answer in standard form.

3

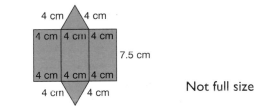

4 cm 4 cm
4 cm 4 cm 4 cm
7.5 cm
4 cm 4 cm 4 cm
4 cm 4 cm
Not full size

> **Comments**
> The net should be drawn, using the fact that the cross-section is an equilateral triangle. The lengths should be drawn to the required accuracy of ±1mm and the angles are drawn to the required accuracy of ±1°.

4 0.067

> **Comments**
> Rounding as $\frac{30 \times 0.1}{15 \times 3} = 0.066\,6666...$

5 The data can be represented as a single stem and leaf diagram as follows.

```
         Article A        Article B
                     2  8 9
         9 9 8 6 2   3  0 1 3 4 6 6 7 8 8 9
9 7 7 5 5 4 4 3 3 1   4  0 1 1 2 2 6
             1 0 0 5
```
Key for A: $9|3 = 39$
    for B: $2|8 = 28$

> **Comments**
> From the stem and leaf diagram you can see that the paragraphs in article A are generally longer than the paragraphs in article B.

6 $y = 5x - 7$

> **Comments**
> Substituting the points (2, 3) and (1, ⁻2) in the equation $y = mx + c$:
> $3 = 2m + c$
> $^-2 = 1m + c$
> The process of elimination is applied to the two equations which are subtracted to produce:
> $5 = m$
> This value is then substituted into the first equation to give $c = ^-7$.

7 **a)** 72 **b)** nth term $= 2n^2$
   c) 50th term $= 5000$

> **Comments**
> **a)** $\quad 2 \quad 8 \quad 18 \quad 32 \quad 50 \quad 72 ...$
> Differences $+6 \; +10 \; +14 \; +18 \; +22$
> **b)** The differences are all multiples of 2. Each term is twice a square number.
> **c)** 50th term $= 2 \times 50^2 = 5000$

8 $2\pi r(h + r)$

> **Comments**
> $2\pi r$ is a common factor and is taken outside the brackets.

9

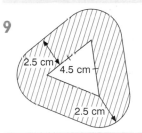

2.5 cm 4.5 cm
2.5 cm
Not full size

> **Comments**
> The locus should be constructed with ruler and compasses, and angles should be measured accurately. The curved areas at the ends are parts of circles.

10 Volume $= 363\pi \, \text{mm}^3$

> **Comments**
> Consider the coin to be a cylinder so that:
> Volume $= \pi r^2 h$
> $= \pi \times 11^2 \times 3$ Where $r = \frac{1}{2} \times$ diameter
> $= 363\pi \, \text{mm}^3$ leaving the answer in terms of π

11 $x = 7$ cm or 2.29 cm (to 2 d.p.)

Comments

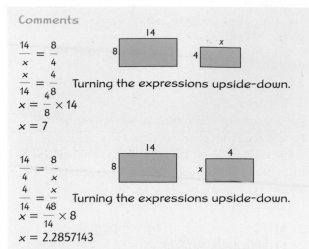

$$\frac{14}{x} = \frac{8}{4}$$

$$\frac{x}{14} = \frac{4}{8}$$ Turning the expressions upside-down.

$$x = \frac{4}{8} \times 14$$

$$x = 7$$

$$\frac{14}{4} = \frac{8}{x}$$

$$\frac{4}{14} = \frac{x}{8}$$ Turning the expressions upside-down.

$$x = \frac{4}{14} \times 8$$

$$x = 2.2857143$$

It is important to round off the answer to a reasonable degree of accuracy and include the units so that $x = 2.29$ cm (3 s.f.).

The way that the question is asked does suggest the possibility of more than one solution as it says 'possible values'. Always take care to check whether a second solution exists.

12 The expression $\frac{4}{3}\pi$ is a constant and r^2 gives units of length $\times$ length $=$ area so the expression cannot represent the volume of a sphere.

13 **a)** 18   **b)** 18

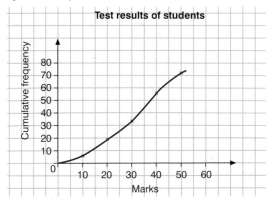

Test results of students

(y-axis: Cumulative frequency; x-axis: Marks)

c) Median = 33 and interquartile range = 21
d) The interquartile range is smaller so the results are less spread out in the second test.

Comments

a) 18   from table
b) 18   72 − 54

Marks	Cumulative frequency	Frequency
⩽ 10	6	6
⩽ 20	19	13
⩽ 30	33	14
⩽ 40	56	23
⩽ 50	72	16

When plotting the cumulative frequency diagram, remember to plot the points at the upper class boundaries (i.e. 10, 20, 30, 40 and 50).

c) Median = 33
Lower quartile = 19
Upper quartile = 40
Interquartile range = 40 − 19 = 21
Always show your working, so that the examiner can see where your answers come from.

d) The conclusions should be stated clearly.

14 $x = 4$ and $y = 3$

Comments

The process of elimination is applied to the two equations which are added to produce:

$7x = 28$ and $x = 4$

This value is then substituted into the first equation to give $y = 3$.

15 **a)**

		red				
		1	2	3	4	5
blue	5	5	10	15	20	25
	6	6	12	18	24	30
	7	7	14	21	28	35
	8	8	16	24	32	40
	9	9	18	27	36	45

b) p(square number) $= \frac{4}{25}$
c) p(final score is less than 30) $= \frac{19}{25}$

Comments

a) Using an appropriate grid helps to show the final scores.
b) Results can be read from the table.
c) Results can be read from the table.

16 **a)** $\angle ROP = 138°$ **b)** $\angle RSP = 69°$
      c) $\angle RQP = 111°$

Comments

a) $\angle OPT = \angle ORT = 90°$

As the tangents to a circle are perpendicular to the radius at the point of contact.

$\angle ROP = 360° - (42° + 90° + 90°)$

As the angles of quadrilateral RTPO add up to 360°.

$\angle ROP = 138°$

b) $\angle RSP = \frac{1}{2} \times \angle ROP$

The angle subtended at the circumference of a circle equals half of the angle subtended at the centre.

c) $\angle RQP = 180° - \angle RSP$

Opposite angles of cyclic quadrilateral RQPS add up to 180°.

$-RQP = 180° - 69° = 111°$

Remember that the diagrams in these questions will not be drawn accurately so you should not attempt to reach solutions by using measuring instruments.

17 a) $3\sqrt{3}$ **b)** $6\sqrt{2}$

Comments

a) The expressions can be simplified using the fact that

$\sqrt{a} \times \sqrt{b} = \sqrt{a \times b}$ and $\frac{\sqrt{a}}{\sqrt{b}} = \sqrt{\frac{a}{b}}$.

$\sqrt{12} + \sqrt{3} = \sqrt{4 \times 3} + \sqrt{3} = 2\sqrt{3} + \sqrt{3} = 3\sqrt{3}$

b) $\sqrt{12} \times \sqrt{6} = \sqrt{12 \times 6} = \sqrt{72} = \sqrt{36 \times 2} = 6\sqrt{2}$

18 $0.2\dot{3}\dot{4} = \frac{26}{111}$

Comments

A rational number can be expressed in the form $\frac{p}{q}$ where p and q are integers.

$1000 \times 0.2\dot{3}\dot{4} = 234.234234...$

$1 \times 0.2\dot{3}\dot{4} = 0.234234...$

$999 \times 0.2\dot{3}\dot{4} = 234$

$0.2\dot{3}\dot{4} = \frac{234}{999} = \frac{26}{111}$ Cancelling down.

As $0.2\dot{3}\dot{4} = \frac{26}{111}$ then $0.2\dot{3}\dot{4}$ must be a rational number.

19

Age group (years)	10–	20–	30–	45–	50–	70–100
Frequency	2	0	6	4	22	12
Class width	10	10	15	5	20	30
Frequency density	0.2	0	0.4	0.8	1.1	0.4

Comments

You need to work out the class width and the frequency density before drawing the histogram.

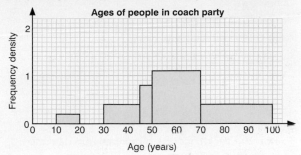

Ages of people in coach party

20 $v = \frac{p}{t-1}$

Comments

$t = \frac{v + p}{v}$

$tv = v + p$ Multiplying both sides by v.

$tv - v = p$ Subtracting v from both sides.

$v(t - 1) = p$ Factorizing the left-hand side.

$v = \frac{p}{t-1}$ Dividing both sides by $(t - 1)$.

21 $\frac{2x - 1}{(x + 3)(x - 4)}$

Comments

A common denominator is $(x + 3) \times (x - 4)$ so writing each part as an equivalent fraction:

$\frac{1}{x + 3} + \frac{1}{x - 4}$

$= \frac{x - 4}{(x + 3)(x - 4)} + \frac{x + 3}{(x + 3)(x - 4)}$ where $\frac{1}{x + 3} = \frac{x - 4}{(x + 3)(x - 4)}$

$= \frac{x - 4 + x + 3}{(x + 3)(x - 4)}$ and $\frac{1}{x - 4} = \frac{x + 3}{(x + 3)(x - 4)}$

$= \frac{2x - 1}{(x + 3)(x - 4)}$

22 a)   **b)**

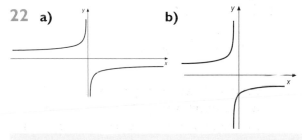

Comments

a) This moves the graph along the x-axis, one unit to the right.

b) This shrinks the graph along the x-axis.

CALCULATOR PAPER

I 24 mph and 36 mph

Comments

Ratio = 2 : 3 : 5 = 24 : 36 : 60, as an equivalent ratio found by multiplying by 12. So the speed in the other two lanes is 24 mph and 36 mph.

2 $\frac{4}{1}$ or 4

Comments

To find the reciprocal of a number you need to convert the number to a fraction.

The number $0.25 = \frac{1}{4}$ and turning the fraction upside-down gives the answer $\frac{4}{1}$ or 4.

3 Not all amounts of money are covered.
No time period is specified for the answer.

Comments

Full answers are essential for maximum marks on this type of question. Similar answers also gain full credit but it is important to be clear exactly what is meant. Other acceptable answers include:

'The classes are too wide for further analysis.'
'At least four classes should be provided in the questionnaire.'
'The respondents may not get the same amount each week.'

4

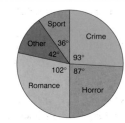

Comments

Always label each sector and include the angles at the centre for further information.

Sum of frequencies = 120 so each book gets $360° ÷ 120 = 3°$.

Angles of pie chart:

Type of book	Frequency	Angle
Sport	12	$12 \times 3° = 36°$
Crime	31	$31 \times 3° = 93°$
Horror	29	$29 \times 3° = 87°$
Romance	34	$34 \times 3° = 102°$
Other	14	$14 \times 3° = 42°$
		Total 360°

Always check that the angles add up to 360°.

5 **a)** 60 kph **b)** 65.5 kph (3 s.f.)

Comments

a) For speed in kilometres per hour, the time must be expressed in hours.

40 minutes = $\frac{40}{60}$ hours

Between A and B distance travelled = 40 km and time taken = 40 minutes.

So speed = distance ÷ time = $40 ÷ \frac{40}{60}$ = 60 kph

b) Again 110 minutes = $\frac{110}{60}$ hours.

Between A and C distance travelled = 120 km and time taken = 110 minutes.

So speed = distance ÷ time = $120 ÷ \frac{110}{60}$
= 65.454 545 = 65.5 kph (3 s.f.)

6 $x > -\frac{1}{2}$

Comments

$2(3 + 2x) > 4$

$6 + 4x > 4$ Expanding the brackets.

$4x > {}^-2$ Subtracting 6 from both sides.

$x > -\frac{1}{2}$ Dividing both sides by 4.

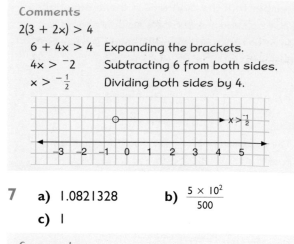

7 **a)** 1.0821328 **b)** $\frac{5 \times 10^2}{500}$

c) 1

Comments

Rounding off to 1 s.f.

8 £534000

Comments

After an increase of 8%, £576720 represents 108% (100% + 8%) of the original value so

108% = £576720, 1% = £5340 (dividing by 108) and 100% = £534000.

9 $x = 2.1$ (1 d.p.)

Comments

Use trial and improvement to home in on the solution.

10 **a)** $A = 5030$ cm^2 (3 s.f.)

b) $r = \sqrt{\dfrac{A}{4\pi}}$

c) $r = 2.82$ cm (3 s.f.)

Comments

a) $A = 4 \times \pi \times r^2 = 5026.5482$ cm^2

b) $A = 4\pi r^2$

$\frac{A}{4\pi} = r^2$ Dividing both sides by 4π.

$r^2 = \frac{A}{4\pi}$ Turning the equation around.

$r = \sqrt{\frac{A}{4\pi}}$ Taking square roots on both sides.

c) $r = \sqrt{\frac{A}{4\pi}} = \sqrt{\frac{100}{4\pi}} = 2.820\,947\,9$ cm

11 $20x$

Comments

$(x + 5)^2 - (x - 5)^2 = \{x^2 + 10x + 25\} - \{x^2 - 10x + 25\}$
$= x^2 + 10x + 25 - x^2 + 10x - 25$
$= 20x$

12 $\angle PBC = 72°$ Exterior angle of regular pentagon.
$\angle PCB = 72°$ Exterior angle of regular pentagon.
$\angle BPC = 36°$ Angles of a triangle add up to 180°.

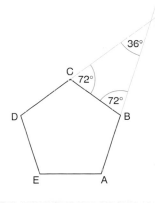

Comments

It is always a good idea to give reasons, so that the examiner can give credit for the methods used, even if the answers are wrong.

External angle of regular pentagon = $\frac{360}{5} = 72°$.
$\angle PBC = \angle PCB = 72°$ as these are exterior angles.
$\angle BPC = 36°$ as the angles of triangle BPC add up to 180°.

13   Area of the washer = 84.8 mm^2 (3 s.f.)

Comments

Calculate the cross-sectional area by subtracting the area of the smaller circle from the area of the larger circle, remembering to halve the diameter to give the radius each time.
Area = $\pi \times 6^2 - \pi \times 3^2 = 84.823\,002$ mm^2

14   Distance = 1330 m (3 s.f.)

Comments

Start by drawing a sketch of the situation and completing the given details on the diagram. Since the angle of depression is 20° the top angle in the triangle is 70° (as angles forming a right angle add up to 90°). Then from the triangle:

$\tan 70° = \frac{x}{485}$

$x = 485 \times \tan 70° = 1332.5265$ m

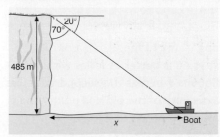

15   **a)** $\angle ABD = 90° - x$ **b)** $\angle DBE = x$
c) $\angle BAD = 90° - 2x$

Comments

a) $\angle CDE = 90°$ CE is a diameter and the angle in a semicircle = 90°.

$\angle DEC = 90° - x$ As the angles of the triangle DEC add up to 180°.

$\angle DBC = 90° + x$ As opposite angles of a cyclic quadrilateral add up to 180°.
$180° - (90° - x) = 90° + x$

$\angle ABD = 90° - x$ As the angles on a straight line ABC add up to 180°.
$180° - (90° + x) = 90° - x$

b) $\angle DBE = DCE$ As the angles subtended by the same chord DE at the circumference are equal (it would be helpful to draw in the line BE on the diagram).

c) $\angle ACE = 3x$ and $\angle AEC = 90° - x$
$\angle BAD = 180° - \{3x + (90° - x)\}$
   As the angles of $\triangle$ ACE add up to 180°.

$\angle BAD = 180° - (3x + 90° - x)$
$= 180° - (2x + 90°) = 180° - 2x - 90°$
$= 90° - 2x$

16 a)

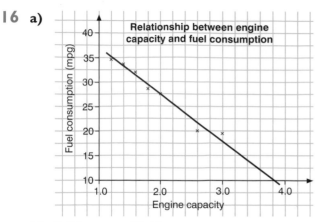

Relationship between engine capacity and fuel consumption

b) (i) 24 mpg   (ii) 13 mpg

c) The first value is better as it lies within the plotted points whereas the second value lies at the extremity of the plotted points and is therefore more prone to error.

Comments

You should use a line of best fit to find estimates for the given questions. More marks are awarded if you provide a thorough explanation for the choice. This should always be borne in mind when undertaking work of this nature.

17 a) £2000

b) 2 years 10 months (to nearest month)

Comments

a) Using $v \propto \frac{1}{a}$ to write $v = \frac{k}{a}$, find the value of k by substituting $v = 7000$ when $a = 1$.

The constant of proportionality = 7000 and $v = \frac{7000}{a}$.

When $a = 3\frac{1}{2}$, $v = £2000$.

b) When $v = 2500$, $a = 2.8$ years.

It is important that this is not interpreted as 2 years 8 months as it is closer to 2 years 10 months (to the nearest month).

18 a)

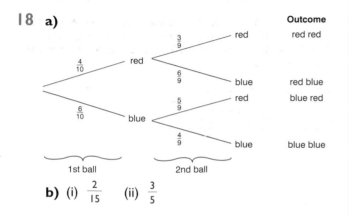

Outcome

$\frac{4}{10}$ red $\frac{3}{9}$ red	red red
$\frac{6}{9}$ blue	red blue
$\frac{6}{10}$ blue $\frac{5}{9}$ red	blue red
$\frac{4}{9}$ blue	blue blue

1st ball   2nd ball

b) (i) $\frac{2}{15}$ (ii) $\frac{3}{5}$

Comments

b) (i) The probability of obtaining two red balls
= p(red and red)
= $\frac{4}{10} \times \frac{3}{9} = \frac{2}{15}$

(ii) The probability of obtaining one ball of each colour
= p(red and blue or blue and red)
= $\frac{4}{10} \times \frac{6}{9} + \frac{6}{10} \times \frac{5}{9} = \frac{54}{90} = \frac{9}{15} = \frac{3}{5}$

19 Area = 1080 cm² (3 s.f.)

Comments

Area = $\frac{110}{360} \times \pi \times 35^2 - \frac{110}{360} \times \pi \times 10^2$
= $\frac{110}{360} \times \pi \times (35^2 - 10^2)$
= 1079.9225

20 $\theta = 0°, 45°, 135°$ and $180°$

Comments

From the sketch, the values of θ which are common to both graphs can clearly be seen.

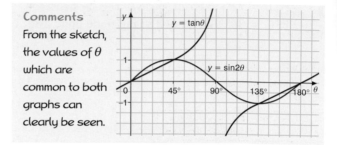

21 a) The angle edge CP makes with the base = 73.1° (3 s.f.)

b) The angle face APD makes with the base = 77.9° (3 s.f.)

c) The area of one of the triangular faces = 10.7cm³ (3 s.f.)

Comments

a) By Pythagoras' theorem:
AC = $\sqrt{3^2 + 3^2}$ = 4.242 640 7
NC = 2.121 320 3
Where N is the centre of the base.

tan NCP = $\frac{7}{2.121\,320}$
∠NCP = 73.140 79°

b) tan θ = $\frac{PM}{NX} = \frac{7}{1.5}$
Where X is the midpoint of AD.
θ = 77.905 243°

c) Using Pythagoras' theorem in triangle PXN:
PX² = PN² + XN²
PX = $\sqrt{7^2 + 1.5^2}$ = 7.158 910 5
Area of triangle
= $\frac{1}{2} \times$ base $\times$ perpendicular height
= $\frac{1}{2} \times 3 \times 7.158\,910\,5$
= 10.738 366 cm³

22 Maximum = 73.9 mph (3 s.f.)

Minimum = 70.1 mph (3 s.f.)

Comments

Distance = 42 miles. Distance_{min} = 41.5 miles.

Distance_{max} = 42.5 miles.

Time = 35 minutes. Time_{min} = 34.5 minutes.

Time_{max} = 35.5 minutes.

Using speed = $\frac{distance}{time}$, and expressing time in hours:

$\text{Speed}_{max} = \frac{42.5}{\frac{34.5}{60}} = 73.9$ mph (3 s.f.)

$\text{Speed}_{min} = \frac{41.5}{\frac{35.5}{60}} = 70.1$ mph (3 s.f.)

ACKNOWLEDGEMENTS

Published by HarperCollins*Publishers* Ltd
77-85 Fulham Palace Road
London W6 8JB

www.CollinsEducation.com
On-line support for schools and colleges

© HarperCollins*Publishers* Ltd 2003

First published 2001
This new edition published 2004.

ISBN 0 00 717089 0

Paul Metcalf asserts the moral right to be identified as the author of this work.

British Library Cataloguing in Publication Data
A catalogue record for this book is available from the British Library.

Edited by Joan Miller
Production by Katie Butler
Series design by Sally Boothroyd
Book design by Ken Vail Graphic Design, Cambridge
Index compiled by Julie Rimington
Printed and bound in Hong Kong by Printing Express Ltd.

Acknowledgements
The Author and Publishers are grateful to the following for permission to reproduce copyright material:

AQA (pp. 177 q7, 178 q17, 179 q19, pp. 179 q20, 180 q3, 181 q9, 182 q16, 183 q5, 183 q21)
Answers to questions taken from past examination papers are entirely the responsibility of the author and have neither been provided nor approved by the AQA. They may not constitute the only possible solutions.

Edexcel Foundation (p. 182 q15)
Edexcel Foundation accepts no responsibility whatsoever for the accuracy or method of working in the answers given.

OCR (pp. 11 q3, 19 q1, 169 q1, 178 q15)
OCR bears no responsibility for the example answers to questions taken from its past question papers which are contained in this publication.

Illustrations
Roger Bastow, Harvey Collins, Gecko Ltd and Tony Warne

Every effort has been made to contact the holders of copyright material, but if any have been inadvertently overlooked, the Publishers will be pleased to make the necessary arrangements at the first opportunity.

You might also like to visit:

www.fireandwater.com
The book lover's website